Advances in Composite and Advanced Materials Machining

This book presents cutting-edge research in effective machining methods for composite materials, including metals, fiber-reinforced polymers, and alloys.

By explaining how to effectively enhance machine life and optimize materials and time costs, *Advances in Composite and Advanced Materials Machining* enables readers to create the best possible end product. Focusing on common modern materials including novel composites, polymers, lightweight alloys, and advanced materials, the book also provides a step-by-step guide to effective machining of hard-to-cut material. It also covers trimming, milling, drilling, and other modern machining processes on fiber-reinforced polymer composites. Recent advances in drilling polymeric matrix composites, ecological machining, grinding technology, nano-machining, and intelligent machining are all covered.

This book will be of interest to professionals in aerospace and automotive engineering as well as unconventional machining, advanced manufacturing processes, and nanomachining.

Advances in Composite and Advanced Materials Machining

Edited by

S.P. Jani, A. Senthil Kumar and
M. Adam Khan

CRC Press
Taylor & Francis Group
Boca Raton London New York

CRC Press is an imprint of the
Taylor & Francis Group, an **informa** business

First edition published 2025
by CRC Press
2385 NW Executive Center Drive, Suite 320, Boca Raton FL 33431

and by CRC Press
4 Park Square, Milton Park, Abingdon, Oxon, OX14 4RN

CRC Press is an imprint of Taylor & Francis Group, LLC

ISBN: 978-1-032-50218-2 (hbk)
ISBN: 978-1-032-50223-6 (pbk)
ISBN: 978-1-003-39746-5 (ebk)

DOI: 10.1201/9781003397465

Typeset in Times
by Apex CoVantage, LLC

Contents

Preface

Machining and machinability play essential roles in the manufacturing industries. Other manufacturing processes produce the parts; often the same parts require further finishing operations before the product is equipped for application. To obtain the desired machinability and performance and enhance machine life, the engineer uses various optimizing techniques for identifying suitable process parameters to achieve a good surface finish, optimizing the material removal rate, and enhancing the tool life.

This book presents modern research and practices for actual and effective machining of hard-to-cut material, providing good technological information on delamination-free trimming, milling, drilling, and other modern machining processes on fiber-reinforced polymer composites.

This book will focus on the machinability of various modern materials such as novel composites, polymers, lightweight alloys, high-temperature alloy materials, and hard materials for modern manufacturing engineering. Moreover, it will give solutions for machining in the aircraft, automotive, defense, aerospace, and other advanced industries.

It aims to provide fundamental knowledge, the latest developments, and research results. This book is intended to aid researchers, engineers, scholars, technical experts, and specialists working in the areas of unconventional machining, advanced manufacturing process, and micro-nanomachining.

Editors

S.P. Jani, PhD, is an Associate Professor in the Department of Mechanical Engineering at Marri Laxman Reddy Institute of Technology and Management, Hyderabad, India. He started to serve as an educator cum researcher in the year 2012. He did his doctorate research on machinability of hybrid fiber/epoxy composites by using abrasive water jet machining at Anna University, Chennai, Tamil Nadu, India. He earned his undergraduate and postgraduate degrees at Anna University, Chennai, India. His research areas include additive manufacturing, advanced manufacturing, machinability study, and machining of natural fiber composites. He is the Editor of *Materials Science and Engineering* for the International Conference on Challenges in Mechanical Engineering (2021), he has written more than 50 papers in various international reputed journals, and he is a reviewer of more than 20 international reputed journals.

A. Senthil Kumar, PhD, is a Professor at Sethu Institute of Technology, Virudhunagar, Tamil Nadu, India, and an administrator and teacher with more than 25 years of teaching and research experience. He was formerly Dean of Mechanical Engineering at Sethu Institute of Technology. He is also an Academic Council member of Anna University of Technology, Tirunelveli. Dr. Kumar has specialized in mechanical engineering and has a flair for research in various fields, including metal machining, ceramic composites, metal matrix composites, and polymer matrix composites. He has published more than 50 technical papers in various international and national journals and 50 technical papers in various reputed international and national conferences. A few of his papers in international journals were the most frequently downloaded articles in 2004 and 2006. Dr. Kumar is also a reviewer in various international journals. He was awarded Outstanding Reviewer by the journal editors of *Materials and Design.*

M. Adam Khan, PhD, is a Professor in the Department of Mechanical Engineering at Kalasalingam Academy of Research and Education, Tamil Nadu, India. He started to serve as an educator cum researcher in the year 2006. He did his postdoctoral research on sustainable manufacturing for advanced materials at the University of Johannesburg, Doornfontein Campus, Johannesburg, South Africa. He earned a PhD at the National Institute of Technology, Trichy, India. He earned undergraduate and postgraduate degrees at Anna University, Chennai, India. His research areas include additive manufacturing, advanced materials, tribology, corrosion, and surface science. Dr. Khan was the Lead Guest Editor for a special issue of the *Journal of Advances in Materials Science and Engineering*. He was the Lead Editor for the book *Innovations in Additive Manufacturing*, Guest Editor for Materials Today Proceedings for the international conference ICLDMM2022, and Guest Editor in *Frontiers in Materials*, Special Call for Energy Storage and Material Metal Oxides. Dr. Khan has published more than 75 technical articles in journals of international repute and more than 50 technical papers in international and national conferences. He holds reviewer status with several leading journals. Dr. Khan has delivered more than 25 lectures at different universities and colleges. His lectures on technical and non-technical topics motivate students towards their career development. He is also a researcher/doctoral committee member for various colleges. His merits are Anna University Third Rank for the Batch 2005 (BE Production Engineering) and Best Reviewer and Outstanding Reviewer status from journal publishers. He is a member of several professional societies, including the Electrochemical Society of India (ECSI) IISc Bangalore.

Contributors

M. Dev Anand
Department of Mechanical
Engineering
Noorul Islam Centre for Higher
Education
Kumaracoil, Kanyakumari District,
Tamil Nadu, India

R. Arun
Department of Mechanical
Engineering
K. Ramakrishnan College of
Engineering
Trichy, Tamil Nadu, India

K.S. Jai Aultrin
Department of Marine Engineering
Noorul Islam Centre for Higher
Education
Kumaracoil, Kanyakumari District,
Tamil Nadu, India

Muthu Chozha Rajan B
Department of Mechanical
Engineering
Sethu Institute of Technology
Kariapatti, Tamil Nadu, India

Jafrey Daniel James D
Department of Mechanical
Engineering
K. Ramakrishnan College of
Engineering
Trichy, Tamil Nadu, India

Kiran M D
Department of Mechanical
Engineering
BMS Institute of Technology and
Management
Bengaluru, Karnataka, India

Udaya Devadiga
Department of Mechanical
Engineering
NMAM Institute of Technology, Nitte
(Deemed to be University)
Nitte, Karnataka, India

U. Elaiyarasan
Department of Mechanical
Engineering
Centre for Additive Manufacturing
Chennai Institute of Technology
Chennai, Tamil Nadu, India

Karthik Pandiyan G
Department of Mechanical
Engineering
Sri Vidya College of Engineering and
Technology
Virudhunagar, Tamil Nadu, India

R.M. Galagali
Department of Mechanical
Engineering
S. G. Balekundri Institute of
Technology
Shivabasav Nagar, Belagavi, India

Ashok M H
Department of Mechanical
Engineering
S G Balekundri Institute of
Technology
Shivabasav Nagar, Belagavi, India

Vishwanath Khadakbhavi
Department of Mechanical
Engineering
S. G. Balekundri Institute of
Technology
Shivabasav Nagar, Belagavi, India

G. Jeeva Kumar
Department of Mechanical
Engineering
K. Ramakrishnan College of
Engineering
Trichy, Tamil Nadu, India

Nithin Kumar
Department of Mechanical Engineering
NMAM Institute of Technology, Nitte
(Deemed to be University)
Nitte, Karnataka, India

P. Senthil Kumar
Department of Mechanical
Engineering
B V Raju Institute of Technology
Narsapur, Telangana, India

Srinivas Prabhu M
Department of Mechanical
Engineering
NMAM Institute of Technology, Nitte
(Deemed to be University)
Nitte, Karnataka, India

Balasundaram R
Department of Mechanical
Engineering
SRM Institute of Science and
Technology
Tiruchirappalli, Tamil Nadu, India

Masi Periyasaame R
Department of Mechanical Engineering
K. Ramakrishnan College of
Engineering
Trichy, Tamil Nadu, India

S. Raja
Department of Mechanical Engineering
Noorul Islam Centre for Higher
Education
Kumaracoil, Kanyakumari District,
Tamil Nadu, India

R. Rajesh
Department of Mechanical Engineering
Noorul Islam Centre for Higher
Education
Kumaracoil, Kanyakumari District,
Tamil Nadu, India

V. Satheeshkumar
Department of Mechanical
Engineering
Government College of Engineering
Salem, Tamil Nadu, India

C. Senthil Kumar
Department of Mechanical
Engineering
University College of Engineering
Panruti, Tamil Nadu, India

Rashmi P Shetty
Department of Robotics and AI
NMAM Institute of Technology, Nitte
(Deemed to be University)
Nitte, Karnataka, India

Sunil Kumar Shetty
Department of Mechanical
Engineering
NMAM Institute of Technology,
Nitte (Deemed to be
University)
Nitte, Karnataka, India

J. Shivakumar
Department of Mechanical
Engineering
Jain College of Engineering
Belagavi, India

Prabaharan T
Department of Mechanical
Engineering
MEPCO Schlenk Engineering
College
Sivakasi, Tamil Nadu, India

1 Influence of Fillers on Machining Process of Polymer Composites

Kiran M D, Nithin Kumar, Rashmi P Shetty,
Sunil Kumar Shetty, and Udaya Devadiga

1.1 INTRODUCTION

In recent years, fibre-reinforced polymers (FRPs) have been emerging materials due to their combination of light weight, extremely high strength, and great stiffness. FRPs are used in a variety of industries, such as the automotive, aerospace, and marine industries, due to their distinct mechanical and physical properties. Polymer matrix composites consist of thermoset/thermoplastic as matrix and particulates/fibres as reinforcements. The behaviour of polymer composites is controlled by the type and properties of the matrix and reinforcements. The properties of polymer composites is largely depending on shape and size of reinforcements (like fabrics, long/short fibres, particles), the composition of matrix and reinforcement. The addition of micro/nanoparticles or fillers as secondary reinforcement in polymer composites plays a significant role in enhancement of the mechanical properties of polymer composites. In addition to mechanical properties, thermal and electrical properties are intrinsic to the filler for biomedical applications, structural applications, and aerospace applications.

The secondary process of polymer composites, such as machining, is a slightly complicated process due to their anisotropic and nonhomogeneous behaviour. These physical characteristics of polymer composites certainly lead to substantial damage of composites, low surface finish quality, and dimensional inaccuracy [1, 2]. It is essential to understand the parameters such as tool geometry/material, depth of cut, and difficulties involved in the machining process due to different orientation of the fibres/fillers of composites [3].

1.2 CONVENTIONAL MACHINING PROCESSES OF POLYMER COMPOSITES

The conventional machining processes, such as turning, milling, and drilling, are used to achieve the required shape and size of polymer composite parts. Many researchers have carried out study on the machining of polymer matrix composites to reduce the defects occurring during the machining process [4]. The

DOI: 10.1201/9781003397465-1

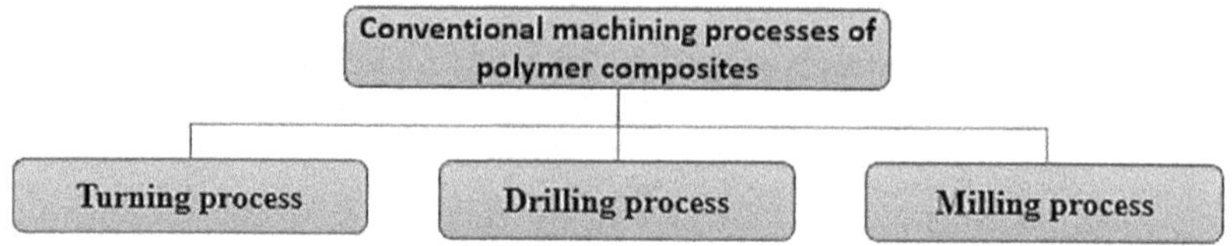

FIGURE 1.1 Conventional machining processes of polymer composites.

machining process of polymer matrix composites mainly depends on material constituents; cutting tool geometry; and machining parameters such as feed, depth of cut, and speed [5].

1.2.1 Influence of Fillers on Turning Process of Polymer Composites

The turning process is the basic conventional machining process used to achieve a cylindrical shape according to dimensions with prescribed tolerances. During the turning process of polymer composites, the feed rate largely affects the surface finish of the product: as the feed rate increases, it causes a poor surface finish on the composite product; however, by increasing the cutting speed during the turning process, an improved surface finish of the composite product will be achieved. The improved surface finish of the composite product can be achieved at low feed rate and high cutting speed during the turning process of polymer composites. The surface finish of the composite product will not be affected much by the depth of cuts during the turning of the polymer composite [6, 7]. During the turning of polymer composites, the areas of chip and cutting forces are associated with each other and directly control the surface finish of the composite product [8].

From several research works on the turning process of polymer composites, researchers identified a few difficulties such as surface integrity, hasty tool wear, and significant damages such as delamination or fibre pull-out. To overcome these difficulties during the machining process, various nano/microfillers or particles such as graphene, carbon nanotubes, silica, and fly ash were used as secondary reinforcement/functional fillers in polymer composites. Polymer composites filled with nanofillers cause a decrease in the damage of composite products during turning, and they exhibit better strength and properties [9]. The feed rate and cutting speeds are the most significant parameters that affect tool wear in uncoated carbides. The surface finish of carbon fibre reinforced polymer (CFRP) products is enhanced at lower cutting force and achieved by inserting coated and uncoated carbides [10].

1.2.2 Influence of Fillers on Drilling Process of Polymer Composites

The drilling process is the most commonly used machining process for polymer composites employed in FRP structural components to join various parts. The drilling process on FRP composites is more difficult than that for conventional materials because a drill tool will pass through different layers of the matrix and reinforcements with different properties. Due to poor surface finish and dimensional imprecision, up to 60% of FRP composite parts are rejected in the assembly process. In addition, fibre pull-out/delamination, cracking, and burning of the fibre and matrix are the major difficulties during drilling of polymer composites. These difficulties cause dimensional imprecision, surface roughness, tool wear, and a decrease in the mechanical strength of the polymer composites. Also, cracking and delamination of polymer composites causes difficulty in achieving dimensional accuracy, axial straightness, a crinkle-free surface, and an accurate circular cross-section. The selection of optimal drilling process parameters such as drill size, depth, speed, and feed attributes can achieve a better drilling process and an excellent surface finish of holes.

Most studies have noticed that high cutting speed with a low feed rate during the drilling process exhibits a better surface finish. During the drilling process of polymer composites, the delamination is higher at the exit of a hole than the entry, and lower delamination can be achieved at a low feed and cutting speed. Poor surface finish and delamination are the most significant defects that occur in drilled polymer composite components. Optimized drilling process parameters will reduce the delamination of composites and enhance the surface finish.

In addition to these, different fillers/particles used as functional fillers or secondary reinforcements minimize or eliminate these drilling process difficulties and improve the strength of the components. Many researchers have proved that the addition of nano/microfillers or particles such as carbon nanotubes, graphene, silicon carbide, fly ash, and boron nitride enhanced the strength and various mechanical properties of polymer composites. In addition, fillers also have effects on the machinability of polymer composites and lead to achieving defect-free drilled components.

The addition of nanographene exhibits better machinability in the drilling process of polymer composites at both the entry and exit of holes. Also, the mechanical properties of jute-reinforced epoxy composites are enhanced with the addition of nanographene. The brittleness of composites increased with the addition of nanographene, and they show higher brittleness at 3 wt.%. The higher brittleness of composites damages the surrounding of hole during drilling operation [11]. CFRP composites filled with 1 wt.% graphene oxide display excellent distribution in the matrix and have better machinability and mechanical properties due to thrust force. And the addition of graphene oxide leads to an increase in the cutting torque and cutting force throughout the drilling process [12, 13].

CFRP composites filled with carbon nanotubes as secondary reinforcements exhibit an improved low delamination factor and drilling tool temperature during the drilling process. Also, the feed rate of the drilling process significantly impacts thrust force and delamination factor. The inclusion of multiwall carbon nanotubes improves the flexural properties of composites and diminishes the thrust force and delamination factor throughout the drilling process. GFRP composites filled with multiwall carbon nanotubes also enhance the fracture toughness along with the minimum delamination factor at hole entry and reduce the edge built up on the hole surface due to the improved thermal properties of composites [14–16].

PA6 composites filled with nanoclay show minimum tool contact at chip interference and thrust force throughout the drilling process due to their brittle nature. The delamination factor throughout the drilling process of polymer composites decreases with the addition of nanoclay and increases at higher wt. % of nanoclay because of poor adhesion and a higher stress concentration [17]. In fly ash-filled CFRP composites, the lower density of fly ash causes deep valley on the hole surface during drilling operation and it leads to poor surface finish of the hole. Also, the addition of fly ash reduces the friction between the material and drilling tool and leads to diminution of thrust force during the process. The addition of fly ash also reduces the delamination factor during the drilling process and shows a minimum value at 10 wt. % fly ash–filled composites [18]. The force required for drilling operation in composites depends on the hardness of the reinforcement/fillers used in the composites. Composites filled with hard particles such as SiC or boron nitride require high force on tools during drilling operation, and it leads to high thrust force. Ceramic particle–filled composites exhibit a low delamination factor and better surface finish during drilling operations due to better interfacial adhesion between the constituents of composites and higher hardness of fillers [19].

1.2.3 INFLUENCE OF FILLERS ON MILLING PROCESS OF POLYMER COMPOSITES

The milling process of polymer composites is a significant and commonly used operation to remove extra/additional materials and is also used to increase the surface finish of the product. Delamination, surface damage, fibre damage/pull-out, and failure of matrix materials are the most common difficulties that occur during milling operation of fibre-reinforced polymer composites due to the properties of the constituents of the composite and machining parameters selected during milling operation. These difficulties in milling operations increase the defects in products and ultimately decrease the life of the product. Many researchers have carried out studies on the machining of fibre-reinforced polymer composites to reduce defects and improve the life of the products by selecting proper machining parameters, constituents of composites, geometry and material cutting tools, fibre orientation, and cutting tool direction. In fibre-reinforced polymer composites, the direction of fibre orientation plays a crucial role, along with tool geometry and its material. Generally, the cutting tool direction is same as the fibre orientation in unidirectional fibres and perpendicular to the fibre orientation in bidirectional/weaving form fibres to minimize the fibre damage/pull-out [20].

Many studies have been done to select the optimum composition of constituents and fillers of composites to minimize product defects during milling operations. Epoxy composites filled with graphene particles minimize the thickness of chips, and it leads to easy cutting operation and chip formation during milling operation due to the inherent lubricant behaviour of graphene. This lubricant property of graphene reduces the wear rate of cutting tools, and also, it requires high cutting force due to the excellent tensile strength and toughness of graphene-filled epoxy composites. The surface accuracy of the epoxy composite product during the milling process is improved by maintaining the feed per number of teeth because of its brittleness. The milling operation of unfilled epoxy composites develops the chips in broken form, and it leads to cracks/cavities on the surface of the product. The addition of graphene to epoxy shows a superior surface finish even at the lower feed per teeth due to toughening mechanism of graphene [21, 22].

MWCNT-filled composites produce thick and continuous chips during milling operations. CFRP composites show the lowest surface roughness and highest cutting force due to better thermal conductivity and strong interference between the CNT and polymer. The polymer phase in CNT-filled composites decides the machining behaviour and shows improved thermal conductivity. Also, the addition of MWCNT reduces the thermal softening effect in the polymer phase and leads to improved machining performance at higher cutting speed [23]. The addition of MWCNT filler with GFRP or polyamide-6 composites reduces the surface roughness and delamination factor due to the lubricant characteristics of MWCNTs at the tool-chip interface, and it reduces matrix debonding and fibre pull-out [24, 25]. Many researchers have carried out studies on the effect of other fillers on milling operations of polymer composites. Nanoclay-filled CFRP composites showed lower surface roughness and delamination factor than unfilled CFRP composites [26].

1.3 CONCLUSIONS

The present chapter focused on the effect of various parameters on the machining (turning, drilling, and milling) process of polymer composites. The constituents used for developing polymer composites can decide the behaviour of composites, and these are effects on the machining properties of polymer composites. Delamination, poor surface finish, surface damage, fibre damage, or pull-out are the most common difficulties that occur during machining of polymer composites. Selecting optimum machining parameters during the machining of polymer composites reduces these difficulties and increases the quality of the products. Selecting proper composition of polymer composites also reduces delamination and fibre damage and improves the surface finish of the composite product. The addition of micro/nanofillers to polymer composites improves their mechanical characteristics because the fillers are stiffer than the polymers. Also, the addition of micro/nanofillers improves the machinability of polymer composites by minimizing machining difficulties such as delamination, poor surface finish, surface damage, and fibre pull-out due to improved bonding between fibre and matrix. The high stiffness, hardness, toughness, and lubricating properties of the fillers also contribute to improving the machinability of polymer composites, reducing defects in the composite product, and improving the life of the component.

REFERENCES

1. Raj SSR, Dhas JER, Jesuthanam C, Challenges on machining characteristics of natural fiber-reinforced composites—a review. J Reinf Plast Compos, 40, 2020. https://doi.org/10.1177/0731684420940773

2. Thakur RK, Singh KK, Ramkumar, J, Delamination analysis and hole quality of hybrid FRP composite using abrasive water jet machining. Mater Today Proc, 33: 5653–5658, 2020. https://doi.org/10.1016/j.matpr.2020.04.056

3. Lopresto V, Caggiano A, Teti R, High performance cutting of fibre reinforced plastic composite materials. Procedia CIRP, 46: 71–82, 2016. https://doi.org/10.1016/j.procir.2016.05.079

4. Che D, Saxena I, Han P, Guo P, Ehmann KF, Machining of carbon fiber reinforced plastics/polymers: A literature review. J Manuf Sci Eng, 136, 2014. https://doi.org/10.1115/1.4026526

5. Vinayagamoorthy R, A review on the machining of fiber reinforced polymeric laminates. J Reinf Plast Compos, 37: 49–59, 2017. https://doi.org/10.1177/073168441773153

6. Rajasekaran T, Palanikumar K, Vinayagam B, Turning CFRP composites with ceramic tool for surface roughness analysis. Procedia Eng, 38, 2012. https://doi.org/10.1016/j.proeng.2012.06.341.

7. Rajasekaran T, Palanikumar K, Arunachalam S, Investigation on the turning parameters for surface roughness using Taguchi analysis. Procedia Eng, 51: 781–790, 2013. https://doi.org/10.1016/j.proeng.2013.01.112.

8. Sauer K, Hertel M, Fickert S, Witt M, Putz M, Cutting parameter study of CFRP machining by turning and turn milling. Procedia CIRP, 88: 457–461, 2020. https://doi.org/10.1016/j.procir.2020.05.079.

9. Kiran MD, Govindaraju HK, Jayaraju T, Evaluation of mechanical properties of glass fiber reinforced epoxy polymer composites with alumina, titanium dioxide and silicon carbide fillers. Mater Today Proc, 5(10): 22421–22424, Part 3, 2018. https://doi.org/10.1016/j.matpr.2018.06.602

10. Abdur Rob SM, Srivastava AK, Turning of carbon fiber reinforced polymer (CFRP) Composites: Process modeling and optimization using Taguchi analysis and multi-objective genetic algorithm. Manuf Lett, 33: 29–40, 2022. https://doi.org/10.1016/j.mfglet.2022.07.012

11. Sridharan V, Raja T, Muthukrishnan N, Study of the effect of matrix, fiber treatment and graphene on delamination by drilling jute/epoxy nanohybrid composite. Arab J Sci Eng, 41: 1883–1894, 2016. https://doi.org/10.1007/s13369-015-2005-2

12. Kumar J, Verma RK, Debnath K, A new approach to control the delamination and thrust force during drilling of polymer nanocomposites reinforced by graphene oxide/carbon fiber. Compos Struct, 253: 112786, 2020. https://doi.org/10.1016/j.compstruct.2020.112786

13. Çelik YH, Kilickap E, Koçyiğit N, Evaluation of drilling performances of nanocomposites reinforced with graphene and graphene oxide. Int J Adv Manuf Technol, 100: 2371–2385, 2019. https://doi.org/10.1007/s00170-018-2875-z

14. Li N, Li Y, Zhou J, He Y, Hao X, Drilling delamination and thermal damage of carbon nanotube/carbon fiber reinforced epoxy composites processed by microwave curing. Int J Mach Tools Manuf, 97: 11–17, 2015. https://doi.org/10.1016/j.ijmachtools.2015.06.005.

15. Karimi ZN, Heidary H, Yousefi J, Sadeghi S, Minak G, Experimental investigation on delamination in nanocomposite drilling. FME Transact, 46: 62–69, 2018. https://doi.org/10.5937/fmet1801062Z

16. Singh KK, Kumar D, Experimental investigation, and modelling of drilling on multi-wall carbon nanotube–embedded epoxy/glass fabric polymeric nanocomposites. Proc Inst Mech Eng B J Eng Manuf, 232: 1943–1959, 2016. https://doi.org/10.1177/0954405416682277

17. Ragunath S, Velmurugan C, Kannan T, Optimization of drilling delamination behavior of GFRP/clay nano-composites using RSM and GRA methods. Fibers Polym, 18: 2400–2409, 2017. https://doi.org/10.1007/s12221-017-7420-4

18. Rajmohan T, Experimental investigation, and optimization of machining parameters in drilling of fly ash-filled carbon fiber reinforced composites. Part Sci Technol, 37: 21–30, 2019. https://doi.org/10.1080/02726351.2016.1205686

19. Premnath AA, Drilling studies on carbon fiber-reinforced nano-SiC particles composites using response surface methodology. Part Sci Technol, 37:478–486, 2018. https://doi.org/10.1080/02726351.2017.1398795

20. Patel P, Chaudhary V, Patel K, Gohil P, Milling of polymer matrix composites: A review. Int J Appl Eng Res, 13(10): 7455–7465, 2018.

21. Arora I, Samuel J, Koratkar N, Experimental investigation of the machinability of epoxy reinforced with graphene platelets. J Manuf Sci Eng, 135: 041007, 2013. https://doi.org/10.1115/MSEC2012-7204

22. Fu G, Huo D, Shyha I, Pancholi K, Alzahrani B, Experimental investigation on micromachining of epoxy/graphene nano platelet nanocomposites. Int J Adv Manuf Technol, 107: 3169–3183, 2020. https://doi.org/10.1007/s00170-020-05190-4

23. Samuel J, Dikshit A, DeVor RE, Kapoor SG, Hsia KJ, Effect of carbon nanotube (CNT) Loading on the thermomechanical properties and the machinability of CNT-reinforced polymer composites. J Manuf Sci Eng, 131: 031008, 2009. https://doi.org/10.1115/1.3123337

24. Zinati RF, Razfar MR, Experimental and modeling investigation of surface roughness in end-milling of polyamide 6/multiwalled carbon nano-tube composite. Int J Adv Manuf Technol, 75: 979–989, 2014. https://doi.org/10.1007/s00170-014-6178-8

25. Sharma D, Singh KK, Thakur RK, Parametric optimization of surface roughness and delamination damage in end milling operation of GFRP laminate modified with MWCNT. Mater Today Proc, 22: 2798–2807, 2020. https://doi.org/10.1016/j.matpr.2020.03.411

26. Kumar K, Singh KK, Thakur RK, Analysis on milling of nanoclay-doped epoxy/carbon laminates using Taguchi approach. In: Yadav S, Singh D, Arora P, Kumar H (eds) Proceedings of international conference in mechanical and energy technology smart innovation, systems and technologies (vol. 174). Springer, 2020. https://doi.org/10.1007/978-981-15-2647-3_49.

2 Machinability and Optimization of Process Parameters on Electrical Discharge Machining of Composite Materials by the MCDM Method

Karthik Pandiyan G, Prabaharan T, Jafrey Daniel James D, and Muthu Chozha Rajan B

2.1 INTRODUCTION

Electrical discharge machining (EDM) is electro-thermal process machining that takes place due to high-frequency electric sparks which are produced by a direct current pulse-generator in which a material is placed inside an dielectric medium. The electric spark produced during the machining process melts and vaporizes the material and forms the final neat shape according to the geometry of the tool. Due to poor machining characteristics and poor surface finish, mass production of materials cannot be used.[1] Due to enhanced properties like high strength at low weight, aluminium composites have been used in the manufacturing sector for varying applications.[2] The addition of magnesium content into the matrix material enhances the wettability of the material, leading to better bonding of the matrix and reinforcements to fabricate Al/AlN-based metal matrix composites (MMCs).[3] In our previous research, silicon carbide (5, 10 and 15 wt.%) was used as reinforcement material in an AA6061-T6 alloy as a matrix. The composite material is fabricated by using stir casting methodology. From the results, it was observed that enhanced properties were achieved when silicon carbide was reinforced at 15 wt.%.[4]

Due to the presence of abrasive reinforcement particles, the tool wear rate was increased on the machining of MMCs, which can be rectified by a non-traditional machining process. An urea-dispersed solution was used as a dielectric medium during the EDM process on machining of MMCs.[5] EDM machining studies were carried out on aluminium reinforced with TiCp composite materials, and the machining process parameters were optimized to achieve a higher MRR and lower Ra with an enhanced surface finish using an ultrasonic vibration tool.[6] EDM

DOI: 10.1201/9781003397465-2

machining studies were carried out on aluminium reinforced with SiCp composite materials, and the machining process parameters were optimized to achieve a higher MRR and with lower Ra by using Taguchi-based grey relational analysis (T-GRA).[7] EDM machining on functionally graded 15–35 vol% of SiC/aluminium alloy–based composite materials and process parameters was optimized to achieve higher MRR increases up to the medium level and decreases and then increases in process parameters.[8] A Taguchi L27 orthogonal array-based experimentation was carried out on friction vibration joining of polypropylene composites. Each factor was studied separately, and ANOVA was utilized for studying the percentage of the contribution.[9] EDM machinability studies were carried out on composite materials, and process parameters were optimized that enhanced the geometric tolerance by using Taguchi-based grey relational analysis.[10]

Machinability studies were carried out by using EDM on composite material, and the process parameters were optimized for sustainable mass production.[11] Machinability studies were carried out by using EDM on $AA7075/Si_3N_4/TiN$ composite material, and the process parameters GRA and RSM were optimized.[12] Machinability studies were carried out by using EDM on $AA7075/Si_3N_4/TiN$ composite material, and the process parameters were optimized to achieve geometric tolerance and enhancement of the form and orientation criteria.[13]

MCDM methods, especially ARAS, were used for the selection of robots based on ranking the alternatives.[14] EDM machining on Si_3N_4-TiN composite material and process parameters were optimized by using the Taguchi-based GRA method to enhance circularity, cylindricity and perpendicularity.[15] PMEDM machining studies were carried out on a nimonic 263 alloy using alumina micropowder as dielectric medium, and the process parameters were investigated to achieve a higher material removal rate.[16] Near-dry EDM machining studies were carried out on AISI M2–grade high-speed steel, and its machining process parameters were optimized by using RSM.[17]

Machinability studies were carried out on mild steel by using a composite material (CUMWCNT)–coated 6061Al electrode as tool material to reduce the tool wear rate.[18] Non-Dominated Sorting Genetic Algorithm-II (NSGA-II), a multi-objective optimization algorithm, and the ANN method were used to optimize the process parameters during the machining of Inconel 718 to achieve maximum MRR.[19] By using a Cu-Ni electrode fabricated via electrodeposition, EDM investigations were performed on polycrystalline diamond material. The Cu-Ni electrode enhanced the MRR by 252% compared to Cu, electrodem MRR increased by 6%, and SR decreased by 15%.[20] Al alloy-based composite materials were fabricated through powder metallurgy (PM) methodology, and EDM studies were carried out to investigate the process parameters for achieving lowered surface roughness.[21]

Machinability studies were carried out on AISI 420 steel using a copper electrode, and the machining process parameters were optimized by using Taguchi-Grey relational analysis to enhance SR and MRR.[22] On LM6/SiC/Dunite composites, WEDM machinability tests were carried out, and the process parameters were optimized by a GRA and ANFIS model, which is used to predict the desired performance characteristics.[23] Electric discharge machining was carried out on a titanium alloy

(Ti-6Al-4V) using copper, aluminium, brass and graphite electrodes. For precision machining, positive tool polarity is used; to achieve rough machining, negative tool polarity is used.[24] In machinability studies, PMEDM was conducted on SKD61 steel, and the process parameters were optimized. The authors also investigated the surface integrity and the effect of the addition of reinforcements.[25]

On 316 L steel, machining studies were carried out by adding nanohydroxy-apatite (HA) particles mixed into the EDM oil, and optimization studies were carried out using the NSGA-II method and ANOVA. A validation error of less than 10% was achieved.[26] Machining studies were conducted on Zircaloy-2 by using graphite, CuW and Cu tool electrodes, and it was optimized by entropy-integrated-grey-VIKOR methods and grey relation analysis.[27]

An investigation was carried out on a liquid nitrogen (LN_2) cryogenic-cooled copper electrode in EDM, and the parameters were optimized by GRA, with a 29% improvement in the average surface achieved.[28] Machining studies were carried out on an Al/10% SiC composite with a conventional electrode and CCEDM process. From the findings, the current was the most dominant factor for surface roughness, with an 18% reduction in EWR.[29] Machining studies were carried out on AISI D2 tool steel by CCEDM and CEDM processes to achieve a 20% reduction in EWR and 19% reduction in Ra.[30] WEDM machining studies were carried out on AA6061-T6/15 wt.15% SiC composites, and the process parameters were optimized by response surface methodology and desirability function analysis.[31]

The ARAS method was utilized for multi-criteria decision making for the evaluation of microclimate in office rooms.[32] The ARAS method was used to solve MCDM problems by investigating qualitative and quantitative information and for different measurement systems.[33] ARAS methodology was used for the application for multiple attribute decision-making problems.[34] A multiple-criteria hierarchy process was implemented with the ARAS method to analyse criteria on various levels.[35]

Thus, from these literature studies, it can be concluded that there are only a few works associated with the machining of MMCs. From many literature studies, it was found that the optimization of EDM of AA6061-T6/15wt.% SiC composites had not yet been studied, and there are no works related to this material. Once fabrication was completed, EDM was carried out with input factors like IP (T_{on}), and a V. L27 orthogonal array of experiments was conducted to employ the output responses; then the output characteristics were studied. Finally, the output was optimized by the ARAS method. Several researchers worked on response surface methodology, GRA and Taguchi techniques only. The current study utilizes ARAS MCDM methodology to determine the best optimal solution to achieve higher MRR and lower TWR, CIR, CYL and PP in the EDM machining process by varying the different IP (T_{on}) and V values considered as input process parameters. MCDM-based ARAS methodology has not yet been studied. The manufacturing sector needs well-established MCDM techniques to determine the best characteristics of responses. By operating machines with the best parameter settings, geometrical tolerance characteristics are enhanced. The MCDM technique of the ARAS method was used to investigate multi-objective optimization. The best results obtained were evaluated using the ARAS utility degree. This resulted in greater MRR and reduced TWR, CIR and CYL as output characteristics. A flow chart for the entropy method weights incorporated in the ARAS method is shown in Figure 2.1.

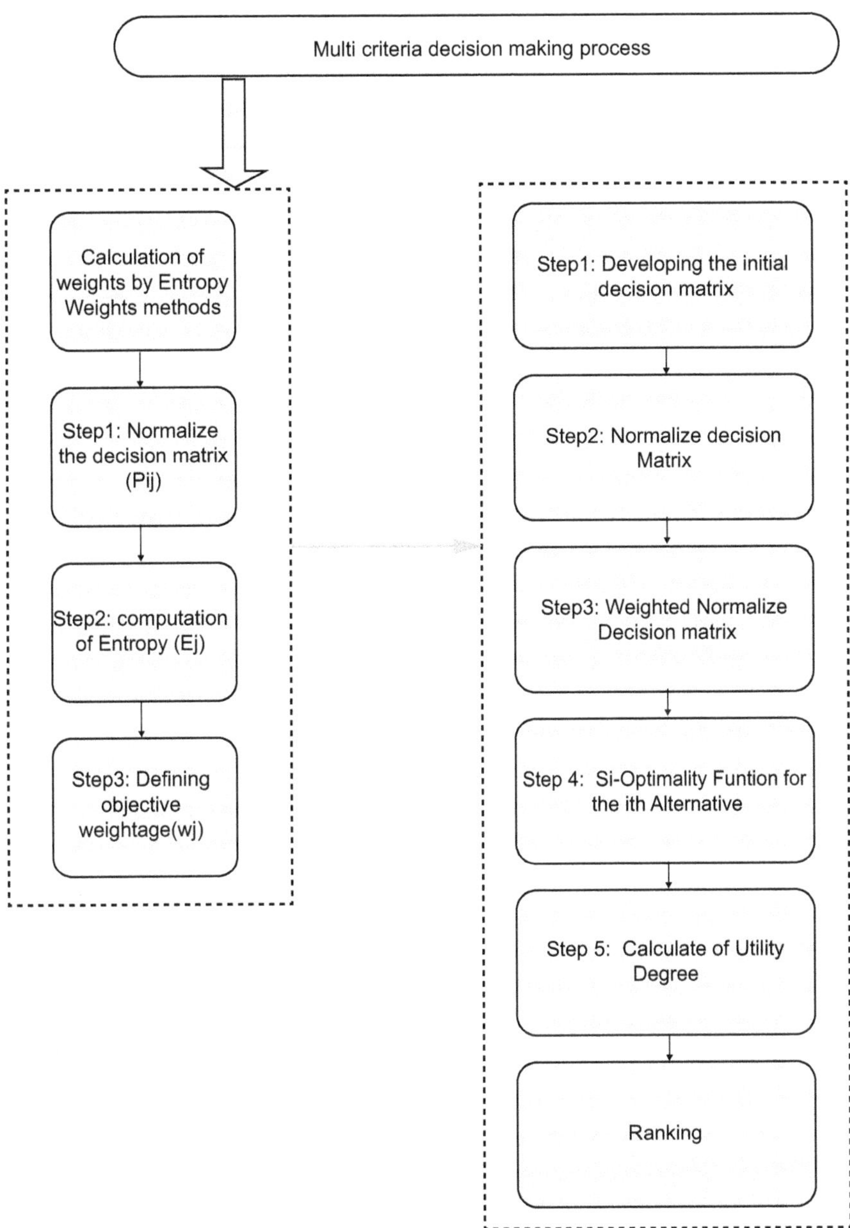

FIGURE 2.1　Flow chart for entropy method weights incorporated in the ARAS method.

2.2 MATERIALS AND METHODS

2.2.1 Fabrication of Material

By using the stir casting methodology, the composite materials were fabricated. An AA6061-T6 alloy was taken as the matrix material. Table 2.1 shows the chemical and physical properties of the AA6061-T6 alloy and AA6061-T6/15 wt.% of SiC composites. The ceramic reinforcement material considered in this study was SiC, which exhibits excellent tribological and mechanical properties. SiC possesses excellent thermal shock resistance, high thermal conductivity and high thermal stability. Hence it was used for the fabrication of the composite material, which was adopted for engine cylinder lining applications. Initially the aluminium alloy was preheated in the furnace at a temperature of 850°C. Ceramic reinforcement particles were preheated to 850°C to eliminate oxidation. AA6061-T6 alloy is initially in the form of a rod that is melted in the casting furnace. Once the metal reaches the liquidus point, a degasser is further added to eliminate the impurities present in the molten metal. After eliminating the impurities, the preheated reinforcement was fed into the molten metal. An ultrasonic agitator with a frequency of 25 KKz was used to stir the molten metal with reinforcement for 10 min. Ultrasonic agitation is an effective method to achieve a uniform homogeneous distribution of reinforcement with the matrix material. One percent wt. of Mg was added into the melt to ensure the proper bonding between the matrix and reinforcement. The homogeneous molten metal was kept at the temperature of 850°C for further processing. The molten metal was poured into a die. Hereafter solidification process of the fabricated composites is taken out from the mould for further machining. Figure 2.2 shows the fabrication of composite materials.

2.2.2 Material Characteristics

Material characterization studies were carried out after the fabrication of composite materials. The tensile strength of the manufactured composites was evaluated according to ASTM B557M. ASTM E9-19 was used to evaluate the compression

---

TABLE 2.1

Chemical and Physical Properties of AA6061-T6 Alloy and AA6061-T6/15 wt.% of SiC Composites

Chemical Properties

Elements in wt. in %	Ti-0.01, Mg-1.08, V-0.01, Fe-0.17, Mn-0.52, Cu-o.32, Si-0.63, Al-97.25

Physical Properties

Tensile strength (MPa)	180 MPa
Density (g/cm³)	2.82 g/cm³
Hardness (HV)	72 BHN

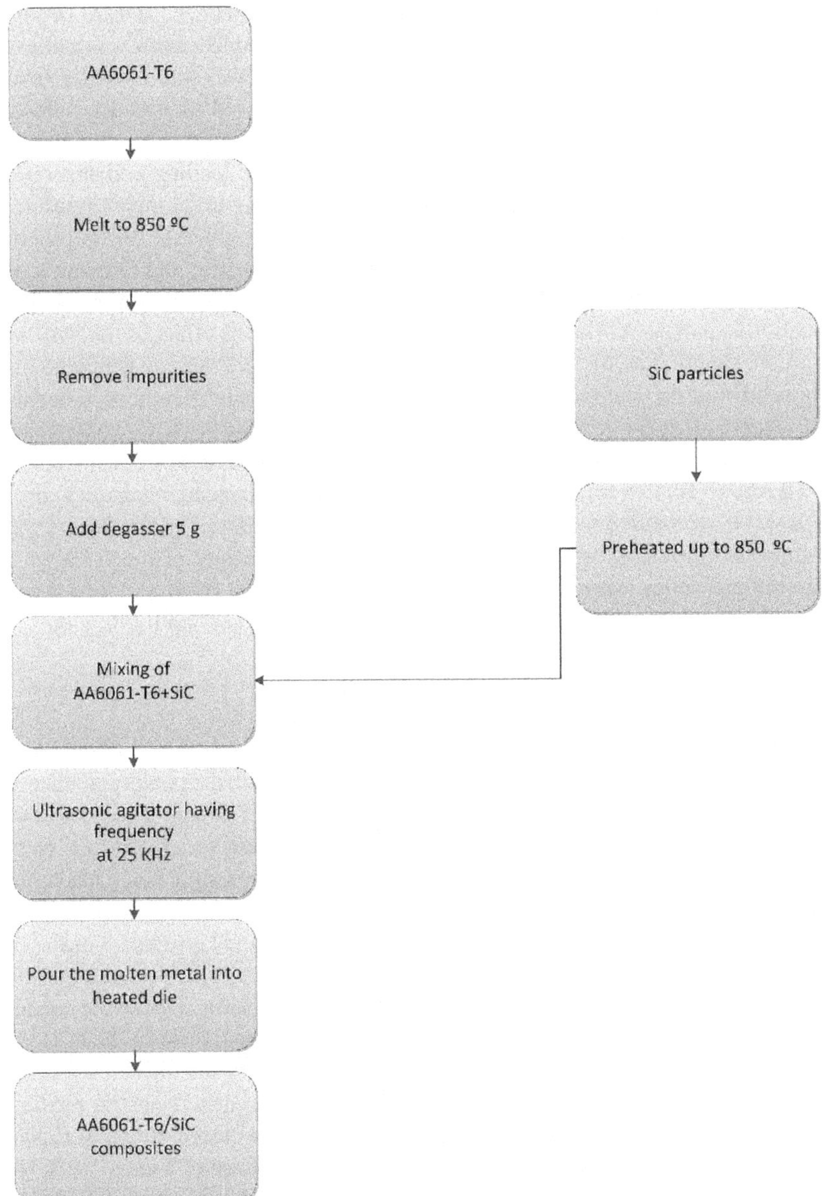

FIGURE 2.2 Fabrication of composite materials.

strength of the manufactured composites and ASTM E23 for the impact strength of the manufactured composites. Brinell hardness analysis was used to measure hardness according to ASTM E10-08 standards. Density measurement was taken according to ASTM D792 standards. Uniform distribution of reinforcements within the matrix achieved good bonding that led to enhanced tensile strength by the Orowan strengthening mechanism. Minor grain dislocation was achieved due to evenly distributed stress during crushing. This was due to better bonding of the matrix as well as its reinforcements, which led to enhanced elasticity that prevented the samples from breaking and enhanced their compressive strength. The addition of reinforcements resisted sudden loading and absorption of higher energy. According to the Orowan strengthening mechanism, reinforcement particles were evenly distributed in the matrix material. Homogeneous distribution of particles restricted the movement of particles, and Orowan loops hindered particle dislocation. Grain boundary strength was enhanced by the dislocation of atoms. While measuring theoretical density, the effect of porosity was not considered. The distance of pouring from the mould causes shrinkage during solidification, leading to the formation of porosity and differences in density. The mechanical properties of the manufactured composites were characterized according to ASTM standards. From the results obtained, it was concluded that the AA6061-T6/15 wt.% of SiC composites exhibited enhanced mechanical properties when compared to the unreinforced AA6061-T6 alloy. The tensile strength, compressive strength, impact strength, hardness and density of the AA6061-T6 unreinforced alloy were 140 MPa, 90 MPa, 27 joules, 60 HRB and 2.64 g/cm^3, whereas they were 188 MPa, 122 MPa, 27 joules, 72 HRB and 2.82 g/cm^3 for AA6061-T6/15 wt.% of SiC composites.[36]

2.2.3 MACHINING PROCESS

After the fabrication of composite materials was completed, the composite material underwent a machining process. A die-sinking type of EDM equipment was used for the machining process. The machine had a three-phase servo stabilizer (servo voltage fixed as 40 units) for maintaining the proper power supply while machining the fabricated composite specimen. The electrode rod used for machining purposes was made of electrolyte copper (length and diameter were 50 mm and 8 mm). The machining workpiece was kept in the die sink machine vice, which acted as positive polarity, whereas the copper electrode acted as negative polarity. Due to the conductivity of the current, the workpiece was weakened and melted, some of the materials evaporated, and then further debris material was cleaned by the dielectric medium (EDM oil). Dielectric conductivity with 20 mho was used as a dielectric medium, which prevented sparks and fire due to the conductivity of current to the workpiece materials while conducting the experiments. The peak voltage of 2 units 110 V DC, servo feed of 2100 units and deionized water with a constant pressure of 15 kg/cm^2 were used in the experiment. The electrode and workpiece were immersed in the dielectric medium while machining. Workpiece material with 8 cm diameter and 3 mm depth was taken from the stir cast AA6061/15%SiC alloy. Table 2.2 shows the experimental details of the EDM process.

Figure 2.3 shows the experimental setup, and Figure 2.4 shows the measure of the output response in CMM.

The most influential input parameters, considered one factor at a time with this study, are *IP* of 12–18 A, T_{on} of 25–75 μs and *V* of 40–50 V for the set of

TABLE 2.2

Experimental Details of EDM Process

Workpiece material	AA6061-T6/1 wt.% SiC (100 × 100 × 10 mm)
Electrode material	Copper (50 mm in length, 8 mm in diameter)
Type of equipment	Die sinking type
Servo voltage	40 V
Dielectric conductivity	20 mho
Current (A)	12, 15, 18 IP
Pulse on time (T_{on})	25, 50, 75 μS
Gap voltage (V)	40, 45, 50 V
Dielectric fluid	EDM oil
Polarity	Work piece (+ve), copper (−ve)
Servo feed	2100 units
Constant pressure	15 kg/cm²
Peak voltage	110 V DC
Gap voltage	0.2 mm

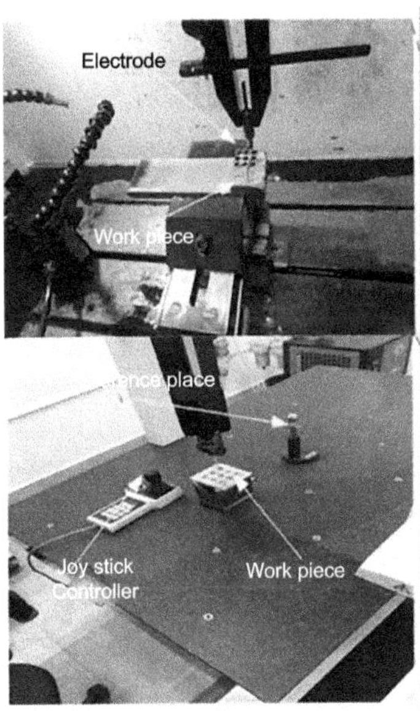

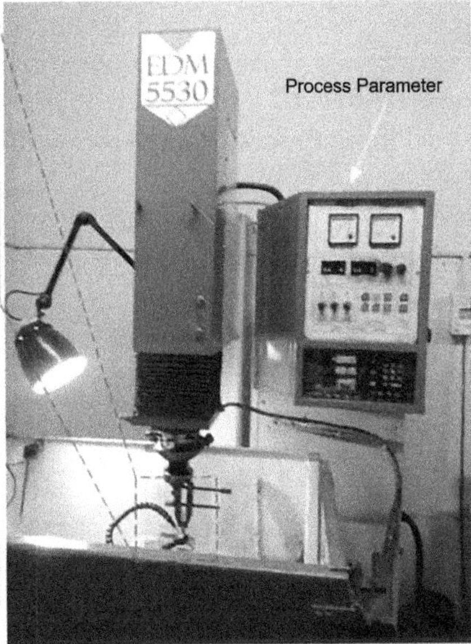

FIGURE 2.3 Experimental setup.

experiments. Output responses like MRR, TWR, CIR, CYL and PP were measured, and their effects on input parameters were investigated. Table 2.3 shows the input process parameters and their levels.

The Taguchi-based L27 orthogonal array of design of experiment methodology was employed for conducting experiments with input process parameters like IP, T_{on} and V. The workpiece and electrode had a gap of 2 mm. The material is

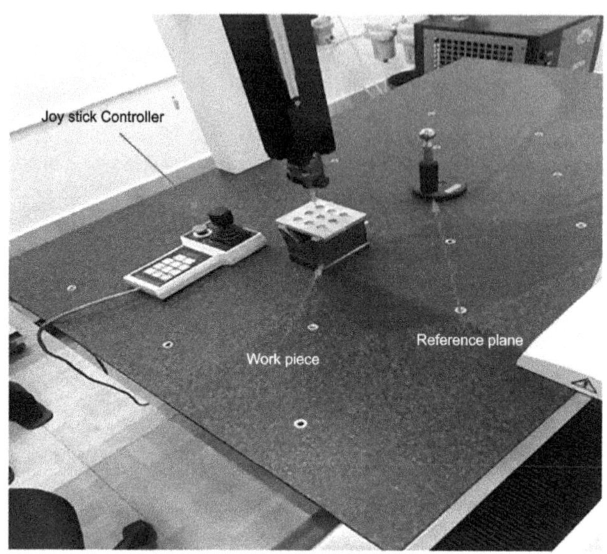

FIGURE 2.4   Measure of output response in CMM.

TABLE 2.3
Input Process Parameters and Their Levels

S. No	Parameters	Level 1	Level 2	Level 3
1	Peak current (*IP*)	12	15	18
2	Pulse on time (μs)	25	50	75
3	Spark gap voltage (*V*)	30	35	40

eroded from the work piece by melting as a result of electrical discharge energy created between the tool and workpiece. The dielectric fluid flushed away the eroded material after the completion of the metal removal process. The eroded debris was stored in the dielectric storage tank. Per the L27 orthogonal array, machining was carried out for all the experiments.

2.2.4 Output Responses

Output responses like MRR, TWR, CIR, CYL and PP, which were obtained from the EDM process parameters, were measured. Output responses like MRR and TWR were evaluated by

$$MRR = Wi - Wf/\rho *t \qquad (1)$$

where *Wi* and *Wf* are the weight of the specimen before and after machining, *t* is the time taken for machining, and ρ is the workpiece material's density =2.820 g/cm^3.

$$TWR = Wi - Wf/\rho \; {}^{*}t \tag{2}$$

where *Wi* and *Wf* are the weight of the tool electrode before and after machining, *t* is the time taken for machining, and ρ is the tool electrode material density = 8.96 g/cm^3.

The other output responses like CIR, CYL and PP were measured utilizing a coordinate measuring machine. The workpiece was placed in a CMM machine, for which the circular plane was designated. A probe was used to measure the CIR, CYL and PP of the fabricated specimen at different positions. CIR, CYL and PP were measured using computer software. It was used to trace the different positions of the machined specimen by an inner and outer layer of the hole machined by the tool. Table 2.4 presents the input parameters with their output responses for all experiments with weight criteria.

2.2.5 OPTIMIZATION BY ENTROPY WEIGHTS INCORPORATED WITH ARAS

The entropy mathematical theory was proposed by C.E. Shannon in 1948. The entropy method is used for the evaluation of weights for MCDM problems and is applied for known data of the decision matrix. There are three major weighting methods: (1) integral weighting methods, (2) subjective weighting and (3) objective weighting. Subjective weighting is further classified into direct weighting and pairwise comparison. Direct weighting includes the simple multi-attribute rating technique (SMART), the SIMOS method, fuzzy individual and group weighting and quality function deployment (QFD). Pairwise comparison includes modified digital logic (MDL), analytic hierarchy process (AHP), eigenvector and the least square method. Objective weighting methods include mean weighting, standard deviation and the entropy method. The entropy method is categorized under the objective weighting method, which is used to evaluate the weights of MCDM problems. Table 2.4 shows the measured output responses for all experiments. To estimate the optimal process, parameters by MCDM were employed by ranking the alternatives with their performance or output responses. Entropy weights are incorporated with the ARAS system to achieve an optimal value of process parameters.

Entropy weighting:

Step 1: Normalize decision matrix array (performance indices) to achieve project outcome (P_{ij}).

TABLE 2.4

Input Parameters with Their Output Responses for All Experiments with Weight Criteria

	Weight Criteria			0.087	0.264	0.244	0.147	0.259
S. No	Current	Pulse on Time	Spark Gap Voltage	MRR	TWR	CIR	CYL	PP
1	12	25	40	0.3411	0.0234	0.0287	0.2043	0.0255
2	12	25	45	0.4421	0.0296	0.0258	0.2567	0.0267
3	12	25	50	0.3740	0.0224	0.0287	0.2325	0.0234
4	12	50	40	0.4592	0.0287	0.0253	0.2114	0.021
5	12	50	45	0.4322	0.0267	0.0260	0.2966	0.0263
6	12	50	50	0.4010	0.0268	0.0244	0.2939	0.0224
7	12	75	40	0.4801	0.0294	0.0253	0.2565	0.0233
8	12	75	45	0.4425	0.0325	0.0289	0.2571	0.0331
9	12	75	50	0.4356	0.0263	0.0224	0.2447	0.0389
10	15	25	40	0.4703	0.0327	0.0320	0.2381	0.0267
11	15	25	45	0.4729	0.0386	0.0285	0.2787	0.0277
12	15	25	50	0.4939	0.0369	0.0322	0.2430	0.0228
13	15	50	40	0.5112	0.0368	0.0255	0.2288	0.0356
14	15	50	45	0.5222	0.0376	0.0365	0.3049	0.0364
15	15	50	50	0.5115	0.0352	0.0478	0.3288	0.0309
16	15	75	40	0.5294	0.0401	0.0433	0.2852	0.0325
17	15	75	45	0.5244	0.0375	0.0311	0.2321	0.0325
18	15	75	50	0.4799	0.0375	0.0310	0.3044	0.0263

19	18	25	40	0.4906	0.0400	0.0287	0.2821	0.03
20	18	25	45	0.5928	0.0425	0.0362	0.3379	0.0316
21	18	25	50	0.5495	0.0481	0.0268	0.3057	0.0394
22	18	50	40	0.5401	0.0426	0.0327	0.3373	0.0365
23	18	50	45	0.5029	0.0458	0.0423	0.3186	0.0392
24	18	50	50	0.4870	0.0424	0.0354	0.2564	0.0422
25	18	75	40	0.5113	0.0359	0.0390	0.3118	0.0398
26	18	75	45	0.5419	0.0450	0.0280	0.3081	0.0378
27	18	75	50	0.5205	0.0454	0.0342	0.3695	0.0334

$$P_{ij} = \frac{x_{ij}}{\sum_{i=1}^{m} x_{ij}} \tag{1}$$

X_{ij}—($i \in \{1, 2, \ldots n\}$ and ($j \in \{1, 2, \ldots m\}$ where the performance value of ith alternative on j^{th} criterion

Step 2: Project outcomes entropy measure.

$$E_j = -K \sum_{i=1}^{m} P_{ij} \ln p_{ij} \tag{2}$$

where $k = 1/\ln (m)$.

Step 3: Objective weight by entropy.

$$W_j = \frac{1 - E_j}{\sum_{j=1}^{n} (1 - E_j)} \tag{3}$$

where W_j = weight criteria.

2.2.5.1 Additive Ratio Assessment Method

MCDM problems deal with the assignment of ranking the decision alternatives (Zavadskas and Turskis, 2010). The ARAS approach is used to determine the utility function value. The utility value determines the best feasible alternative or best optimal parameter, which is based on the output parameter with weight criteria selected (Zavadskas *et al.*, 2008). In the ARAS method, the formulation of the decision matrix is the initial stage. Any discrete optimization problems on MCDM can be solved by initializing the decision matrix, which is represented as follows (Ghram and Frikha, 2019)

Step 1:

$$X = \begin{bmatrix} x_0 & x_{0j} & \cdots & x_{0n} \\ x_{i1} & x_{ij} & \cdots & x_{in} \\ \vdots & \vdots & \vdots & \vdots \\ x_{m1} & x_{mj} & \cdots & x_{mn} \end{bmatrix}, i = \overline{0, m}; j = \overline{1, n} \tag{4}$$

$i = 1, m, j = 1, n$ (n = criteria and m = alternatives),
$X_{ij} = i^{th}$ alternative in terms of the jth criterion performance value,
$X_{0j} = j^{th}$ criterion optimal value.

If j criterion, the optimal value is unknown.

$$x_{0j} = \max_i x_{ij}, \text{ if } \max_i x_{ij} \text{ is preferable;}$$

$$x_{0j} = \min_i x_{ij}^{*}, \text{ if } \min_i x_{ij}^{*} \text{ is preferable.} \tag{5}$$

where x_{ij} = values of performances, and w_j = weight criteria.

Step 2: Normalize all initial criteria values and define the normalized decision matrix $\overline{x}_{ij}$.

$$\overline{X} = \begin{bmatrix} \overline{x}_{01} & \cdots & \overline{x}_{0j} & \cdots & \overline{x}_{0n} \\ \vdots & \ddots & \vdots & \ddots & \vdots \\ \overline{x}_{i1} & \cdots & \overline{x}_{ij} & \cdots & \overline{x}_{in} \\ \vdots & \ddots & \vdots & \ddots & \vdots \\ \overline{x}_{m1} & \cdots & \overline{x}_{mj} & \cdots & \overline{x}_{mn} \end{bmatrix} ; i = \overline{0,m}; j = \overline{1,n} \tag{6}$$

The normalized maximum preferable criteria (for beneficial attributes):

$$\overline{x}_{ij} = \frac{x_{ij}}{\sum_{i=0}^{m} x_{ij}} \tag{7}$$

The minimum preferable criteria values (for nonbeneficial attributes) are normalized in two stages:

$$x_{ij} = \frac{1}{x_{ij}^{*}}; \overline{x}_{ij} = \frac{x_{ij}}{\sum_{i=0}^{m} x_{ij}} \tag{8}$$

Step 3: Normalize the weighted matrix used to estimate the weight for the criteria; the weightage value lies between 0 and 1 (i.e., $0 < w_j < 1$). The sum of all weight criteria must be equal to ($w_j = 1$), which is expressed as follows:

$$\sum_{j=1}^{n} W_j = 1 \tag{9}$$

$$\hat{X} = \begin{bmatrix} \hat{x}_{01} & \cdots & \hat{x}_{0j} & \cdots & \hat{x}_{0n} \\ \vdots & \ddots & \vdots & \ddots & \vdots \\ \hat{x}_{i1} & \cdots & \hat{x}_{ij} & \cdots & \hat{x}_{in} \\ \vdots & \ddots & \vdots & \ddots & \vdots \\ \hat{x}_{m1} & \cdots & \hat{x}_{mj} & \cdots & \hat{x}_{mn} \end{bmatrix} ; i = \overline{0,m}; j = \overline{1,n}. \tag{10}$$

Weighted normalized values of all the criteria:

$$\widehat{x_{ij}} = \overline{x_{ij}}w_j; i = \overline{0,m} \tag{11}$$

where $\overline{x_{ij}}$ is the normalized value of the j criterion, and W_j is the weight of the j criterion.

Step 4: Determine the optimality function by

$$S_i = \sum_{j=1}^{n} \widehat{x_{ij}}; i = \overline{0,m}, \tag{12}$$

where S_i is the ith alternative optimality function.

The higher S_i value is the best value, and the lower value is the worst value.

Step 5: The utility degree K_i is obtained by:

$$K_i = \frac{S_i}{S_o} \tag{13}$$

where S_i and S_0 are the optimality criteria.

The S_i optimality function is directly proportional to the values of x_{ij} and w_j of the criteria. A higher S_i is the most feasible alternative or attribute. The priorities were given per the S_i value to determine the rank for alternatives. The degree of utility is the ratio of the S_i value to the S_o value. S_o is the summation of all the weighted decision matrix values for the ideal best value (Alinezhad and Khalili, 2019). From the results, the estimated values of K_i are between 0 and 1. The optimal parameters for the EDM machined samples with the optimal alternative can be obtained corresponding to the utility degree.

2.3 RESULTS AND DISCUSSION

2.3.1 MICROSTRUCTURE OF COMPOSITE MATERIAL

2.3.1.1 Microstructure Examination

Figure 2.5 shows the micrographs of a matrix of stir cast AA6061-T6 with the addition of 15% SiC composite particles in the (a) polished condition and (b) etched condition.

The dark particles are composite SiC, and the bright zone is the polished metal matrix of AA6061-T6. For the etched condition, Keller's reagent was used as an etchant. The grains of primary aluminium are finer and represent the different rates of cooling of the casting. The composite SiC particles are present both at the grain boundary as well as inside the matrix. The microstructure shows the precipitated Mg_2Si particles in the primary aluminium solid solution. The grain boundaries also show the trapped fine composite particles of SiC. The higher addition of 15% SiC composite particles dense distribution at the grain boundaries with

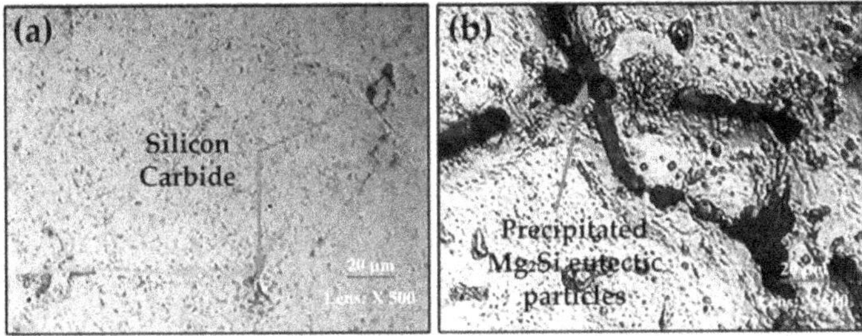

FIGURE 2.5 Micrographs of a matrix of stir cast AA6061-T6 with the addition of 15% SiC composite particles in the (a) polished condition and (b) etched condition.

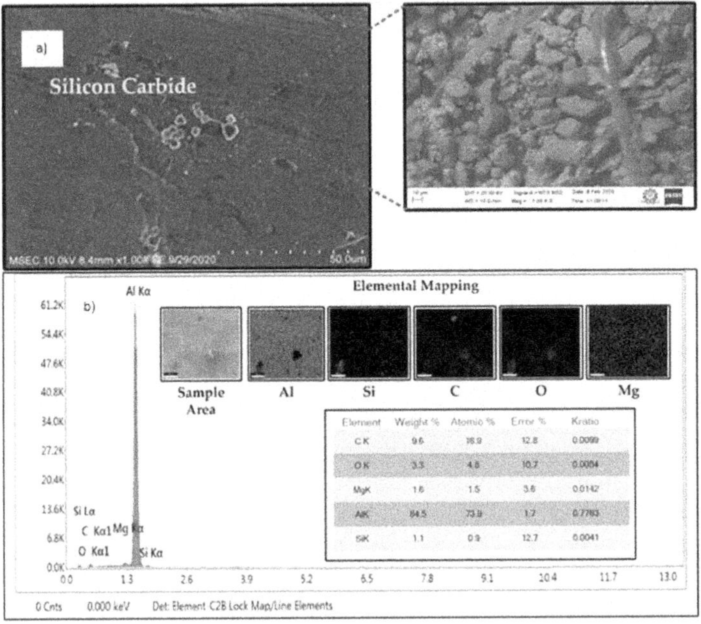

FIGURE 2.6 (a) SEM image of AA6061-T6/15% SiC composite; (b) SEM with elemental mapping of AA6061-T6/15 wt.% SiC composite.

uniform distribution of trapped SiC in the primary aluminium matrix. The grain boundaries are resolved, and the presence of SiC is evident. The precipitated Mg_2Si eutectic particles are also present along the grain boundaries resolved by the primary aluminium grains, grain boundaries, eutectic particles Mg_2Si and composite SiC particles which represents the resolved SiC particles and eutectic particles.

Figure 2.6(a,b) shows the microstructural examination carried out on the fabricated specimen through a scanning electron microscope.

From Figure 2.6(a), it is observed that the microstructure consists of AA6061-T6 alloy with interfacial bonding with eutectic aluminium–silicon atoms. The ceramic reinforcement particles were uniformly dispersed homogeneously with the matrix material, and no porosity was found on the fabricated specimen, which is confirmed by the scanning electron microscope (SEM) investigation. Figure 2.6(b) shows the energy-dispersive analysis carried out on the fabricated specimen. From the EDS analysis, high-intensity peaks of aluminium peaks were found, whereas the other peaks like silicon, oxide and carbon exhibit low intensity. The AA6061-T6 with Mg content was found at the low-intensity peak. The EDS analysis also confirms that reinforcement such as Si and C was found with a strong intensity peak. The secondary reinforcement particles were evenly dispersed in the matrix by the stir casting process during the fabrication of the composite material, which is confirmed by EDS analysis. Figure 2.6(b) shows the SEM with elemental mapping of the AA 6061-T6/SiC composite. The EDS analysis reveals that elements like Al, Si, C and Mg were found in higher amounts in the constituent base metal. From the spectrum, it is shown that the Si and C have been incorporated with the matrix material (AA6061-T6). From EDS analysis, it is observed that low-, medium- and high-intensity peaks were present with the secondary particles. The colours like green, red, yellow, blue and purple represent the overlay mapping of elements present in the composite materials.

2.3.2 ADDITIVE RATIO ASSESSMENT MODEL

Table 2.5 shows the normalized decision matrix for all the experiments. Table 2.6 shows the weighted normalized matrix and optimality function for all the output responses. Table 2.7 shows the utility degree with rankings for all the output responses. The higher utility degree was reflected in the excellent output responses for all the experiments. From ARAS utility degree, it is observed that input parameters like IP at 12 A, T_{on} at 50 μs and V at 40 V generated enhanced output performance characteristics among all the experiments. Table 2.7 shows the ARAS utility degree for all the experimental runs. From Table 2.7, it is found that experiment 4 shows a higher ARAS utility degree of 0.8883.

Table 2.5 shows the value of the normalized decision matrix for all experiments, which is estimated by equations 4 and 5. Table 2.6 shows the weighted normalized matrix for all experiments, which is estimated by equations 6, 7 and 8. Normalized weighted criteria were calculated by equations 9, 10 and 11, and the optimality function was calculated by equation 12. Table 2.7 shows the utility degree with rankings, which is estimated by equation 13.

The ranking of alternatives, according to the values of utility degree, is A3 < A7 < A2 < A1 < A6 < A4 < A5 < A12 < A8 < A10 < A13 < A9 < A11 < A20 < A21 < A22 < A15 < A14 < A16 < A19 < A18 < A23 < A27 < A25 < A24 < A17 < A26. Therefore, A4 is the best optimal parameter corresponding to the utility degree of ARAS methodology.

TABLE 2.5
The Value of the Normalized Decision Matrix for All Experiments

Exp. No	Normalized Decision Matrix				
	MRR	TWR	CIR	CYL	PP
1	0.0250	0.0513	0.0372	0.0471	0.0413
2	0.0324	0.0406	0.0414	0.0375	0.0394
3	0.0274	0.0536	0.0372	0.0414	0.0450
4	0.0336	0.0419	0.0422	0.0455	0.0502
5	0.0317	0.0450	0.0410	0.0324	0.0400
6	0.0294	0.0448	0.0438	0.0327	0.0470
7	0.0352	0.0409	0.0422	0.0375	0.0452
8	0.0324	0.0370	0.0370	0.0374	0.0318
9	0.0319	0.0457	0.0477	0.0393	0.0271
10	0.0344	0.0367	0.0334	0.0404	0.0394
11	0.0346	0.0311	0.0375	0.0345	0.0380
12	0.0362	0.0326	0.0332	0.0396	0.0462
13	0.0374	0.0326	0.0419	0.0420	0.0296
14	0.0382	0.0319	0.0293	0.0315	0.0289
15	0.0375	0.0341	0.0223	0.0292	0.0341
16	0.0388	0.0300	0.0247	0.0337	0.0324
17	0.0384	0.0320	0.0344	0.0414	0.0324
18	0.0352	0.0320	0.0345	0.0316	0.0400
19	0.0359	0.0300	0.0372	0.0341	0.0351
20	0.0434	0.0283	0.0295	0.0285	0.0333
21	0.0402	0.0250	0.0399	0.0315	0.0267
22	0.0396	0.0282	0.0327	0.0285	0.0289
23	0.0368	0.0262	0.0253	0.0302	0.0269
24	0.0357	0.0283	0.0302	0.0375	0.0250
25	0.0374	0.0335	0.0274	0.0308	0.0265
26	0.0397	0.0267	0.0382	0.0312	0.0279
27	0.0381	0.0265	0.0312	0.0260	0.0315

2.3.3 INFLUENCE OF INPUT PROCESS PARAMETERS ON OUTPUT PERFORMANCE CHARACTERISTICS

2.3.3.1 Material Removal Rate

The effect of MRR, TWR, CIR, CYL and PP as input parameters is represented in Figures 2.7 through 2.11.

The material removal rate increases with an increase in the current (IP). A higher amount of discharge energy is produced during the process, leading to an increase in the removal rate of the material. The material rate is increased by increasing the pulse on time (µs). At an elevated pulse on time, a higher amount of

TABLE 2.6

The Weighted Normalized Matrix and the Optimality Function

Exp. No	MRR	TWR	CR	CY	PP	S_i	K_i
1	0.0022	0.0136	0.0091	0.0069	0.0107	0.0424	0.8576
2	0.0028	0.0107	0.0101	0.0055	0.0102	0.0393	0.7954
3	0.0024	0.0142	0.0091	0.0061	0.0116	0.0433	0.8765
4	0.0029	0.0111	0.0103	0.0067	0.0130	0.0439	0.8883
5	0.0027	0.0119	0.0100	0.0048	0.0104	0.0397	0.8040
6	0.0025	0.0119	0.0107	0.0048	0.0122	0.0420	0.8500
7	0.0030	0.0108	0.0103	0.0055	0.0117	0.0413	0.8360
8	0.0028	0.0098	0.0090	0.0055	0.0082	0.0353	0.7141
9	0.0028	0.0121	0.0116	0.0058	0.0070	0.0392	0.7935
10	0.0030	0.0097	0.0081	0.0059	0.0102	0.0370	0.7476
11	0.0030	0.0082	0.0091	0.0051	0.0098	0.0353	0.7132
12	0.0031	0.0086	0.0081	0.0058	0.0119	0.0376	0.7601
13	0.0032	0.0086	0.0102	0.0062	0.0077	0.0359	0.7262
14	0.0033	0.0084	0.0071	0.0046	0.0075	0.0310	0.6271
15	0.0032	0.0090	0.0054	0.0043	0.0088	0.0308	0.6234
16	0.0034	0.0079	0.0060	0.0049	0.0084	0.0306	0.6193
17	0.0033	0.0085	0.0084	0.0061	0.0084	0.0346	0.7004
18	0.0030	0.0085	0.0084	0.0046	0.0104	0.0349	0.7060
19	0.0031	0.0079	0.0091	0.0050	0.0091	0.0342	0.6919

Exp. No	MRR	TWR	CR	CY	PP	S_i	K_i
20	0.0038	0.0075	0.0072	0.0042	0.0086	0.0312	0.6315
21	0.0035	0.0066	0.0097	0.0046	0.0069	0.0313	0.6337
22	0.0034	0.0075	0.0080	0.0042	0.0075	0.0305	0.6167
23	0.0032	0.0069	0.0062	0.0044	0.0069	0.0277	0.5594
24	0.0031	0.0075	0.0074	0.0055	0.0065	0.0299	0.6046
25	0.0032	0.0088	0.0067	0.0045	0.0068	0.0301	0.6095
26	0.0034	0.0071	0.0093	0.0046	0.0072	0.0316	0.6386
27	0.0033	0.0070	0.0076	0.0038	0.0082	0.0299	0.6045

TABLE 2.7
Utility Degree with Rankings

Utility Degree	Ranking
0.8576	3
0.7954	7
0.8765	2
0.8883	1
0.8040	6
0.8500	4
0.8360	5
0.7141	12
0.7935	8
0.7476	10
0.7132	13
0.7601	9
0.7262	11
0.6271	20
0.6234	21
0.6193	22
0.7004	15
0.7060	14
0.6919	16
0.6315	19
0.6337	18
0.6167	23
0.5594	27
0.6046	25
0.6095	24
0.6386	17
0.6045	26

discharge energy was produced between the workpiece and tool, which allowed the material to vaporize and melt from the workpiece, which led to an increase of the removal rate of the material. MRR improves with an increase in gap voltage from a lower to medium level, and then it diminishes as the gap voltage reaches a higher level. The gap voltage increases with a decrease in the material removal rate (Selvarajan, Narayanan and Jeyapaul, 2016). At higher gap voltage, a high amount of discharge energy leads to a crater on the workpiece material and a lower material removal rate. The MRR is enhanced by enhancing the peak current and pulse on time. Combining the effect of current and voltage, the material removal rate is increased by increasing the current, and MRR increases with an increase in gap voltage from lower level to medium level and then diminishes at the higher level (Teng *et al.*, 2020). By a combination of both pulse on time and

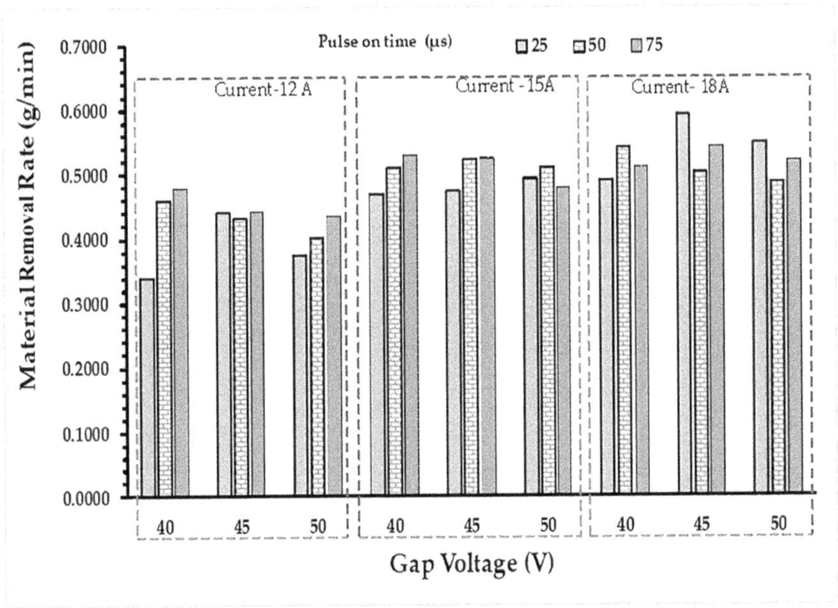

FIGURE 2.7 Influence of input parameters on MRR.

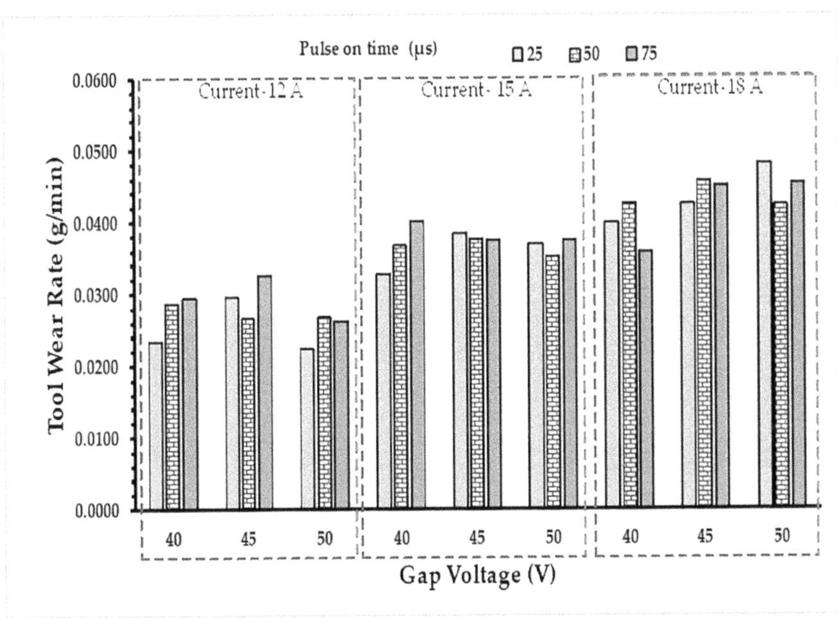

FIGURE 2.8 Influence of input parameters on TWR.

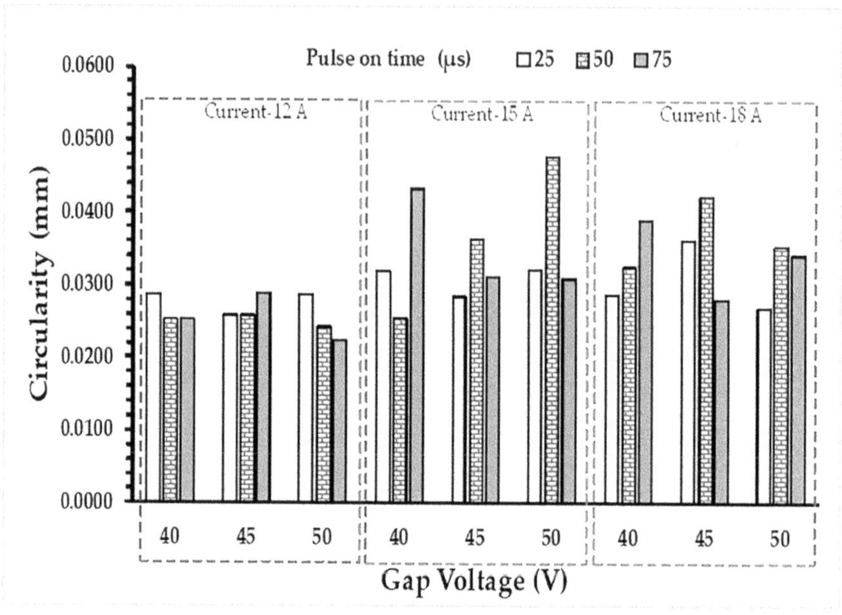

FIGURE 2.9 Influence of input parameters on CIR.

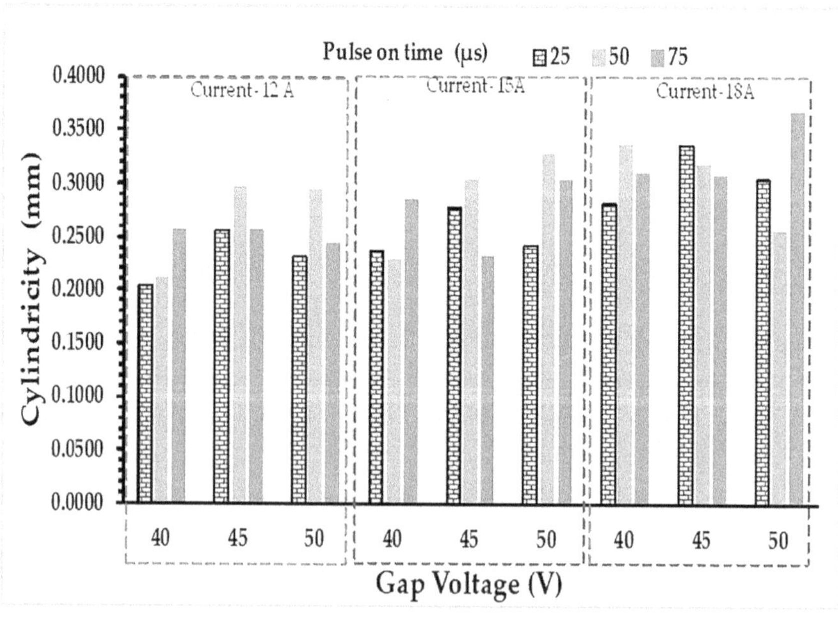

FIGURE 2.10 Influence of input parameters on CYL.

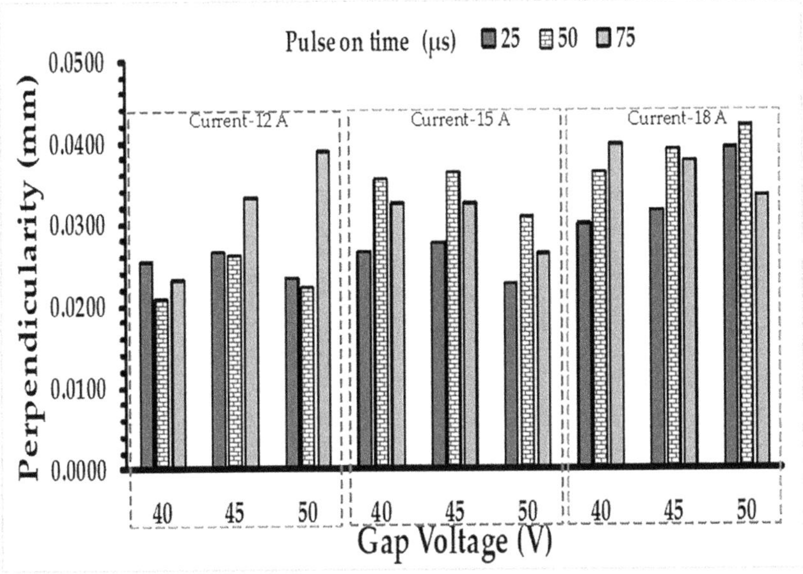

FIGURE 2.11   Influence of input parameters on PP.

gap voltage, the removal rate of the material increases with an increase in pulse on time from a lower level to a higher level, and the removal rate of the material increases with an increase in gap voltage from a lower level to medium level and diminishes at the higher level (Yadav, Kumar and Dvivedi, 2019). With the combination of current and gap voltage, the removal rate of the material increases with an increase in current from a lower level to a higher level, and the removal rate of the material increases with an increase in voltage from a lower level to a medium level and diminishes at a higher level.

2.3.3.2   Tool Wear Rate

Tool wear rate increases with an enhancement in the current (*IP*). A higher amount of discharge energy produced during the process enhances the material rate. The tool wear rate decreases with an increase in pulse on time (μs) from a lower to medium level (Mandal and Mondal, 2019). At a higher pulse on time, a higher amount of thermomechanical discharge energy allows the material get vaporize and melt from the workpiece, which leads to enhancing in TWR. TWR increases with an increase in gap voltage from a lower to medium level and then diminishes as the gap voltage reaches a higher level (Teng *et al.*, 2020). A greater gap voltage results in a higher quantity of discharge energy, which causes a crater on the workpiece material and reduced TWR. This gap voltage is increased when the material removal rate decreases. The TWR increases when the current increases from a lower to a higher level and the TWR increases from a lower to a medium level, after which the improvement in the pulse on time decreases.

Combination of current and voltage, the TWR enhance with enhancement in current, and TWR enhances with enhancement in gap voltage from the lower level to the medium level and then diminishes at the higher level (Sarmah *et al.*, 2020). The TWR increases as the gap voltage increases from a lower to medium level with the combination of both pulse on time and gap voltage. Subsequently, the TWR decreases as the gap voltage increases from a lower to medium level and reaches a higher level. For the combination of current and gap voltage, the tool rate increases with an increase in pulse on time from a lower level to medium level and then diminishes at a higher level, and TWR increases with an increase in voltage from a lower level to medium level and diminishes at a higher level (Kumar S and Kumar M, 2015).

2.3.3.3 Circularity

Circularity increases by enhancing the current (*IP*) up to a lower to medium level and diminishes at a higher level. A higher amount of discharge energy produced during the process leads to greater circularity. Circularity increases with an increase in pulse on time (µs). At a higher pulse on time, a greater quantity of discharge was generated between the tool and the workpiece, leading the material to vaporize and melt from the workpiece. This results in an improvement in circularity (Kumar S and Kumar M, 2015). Circularity diminishes with an increase in gap voltage from lower to higher levels. For the combination of both current and pulse on time, the circularity increases with an increase in current and pulse on time. The circularity of the current–voltage combination increases as the current increases from a low to medium level, and it decreases as the current and voltage increase. Circularity decreases with an increase in gap voltage from a lower level to higher level (Kumar S and Kumar M, 2014). By combining pulse on time and gap voltage, circularity increases as pulse on time increases from a lower to a medium level, and then decreases as pulse on time increases. Similarly, circularity increases as gap voltage increases from a lower to a medium level, and then decreases at the higher level. For the combination of current and gap voltage, the circularity increases with an increase in pulse on time and gap voltage from a lower level to medium level and then diminishes at the higher level (Ahmed *et al.*, 2019).

2.3.3.4 Cylindricity

Cylindricity increases as the current (IP) and pulse on time increase up to a lower to medium level, and it decreases at a higher level. A higher amount of discharge energy produced during the process leads to an increase in cylindricity. A greater quantity of energy was generated between the tool and the workpiece at a higher pulse on time, leading the material to vaporize and melt from the workpiece. This resulted in an increase in cylindricity (Tao Le, 2021). Cylindricity increases with an increase in gap voltage from lower to higher levels. For the combination of both current and pulse on time, the cylindricity increases with an increase in current and pulse on time from the lower level to medium level and then diminishes at the higher level. For the combination of current and voltage, the cylindricity increases with an increase from the lower to medium level and diminishes with an increase in current and voltage at a higher level. Cylindricity increases with an increase in gap voltage from a lower level to higher level (Kumar. S and Kumar. M, 2014). For the combination of both pulse on time and gap voltage, cylindricity increases with an increase in pulse on time and voltage from a lower level to medium level and then diminishes with an increase at a higher level. Cylindricity increases with an increase in gap voltage from a lower level to higher level. For a combination of current and gap voltage, cylindricity increases with an increase in current and gap voltage from a lower level to medium level and then diminishes at the higher level (Kumar S and Kumar M, 2015).

2.3.3.5 Perpendicularity

Perpendicularity increases with an increase in IP. A higher amount of discharge energy is produced during the process and leads to an increase in PP. PP increases with an increase in pulse on time (µs). PP enhances with an increase in gap voltage from a lower to medium level, and then it diminishes with an increase in the gap voltage at a higher level. Gap voltage increases with a decrease in PP (Vinoth Kumar and Pradeep Kumar, 2015). At higher gap voltages, a high amount of discharge energy is produced between the tool and workpiece, which leads to a crater on the workpiece material and lower PP. For the combination of both current and pulse on time, the PP increases with an increase in current and pulse on time (Al-Amin *et al.*, 2021). For the combination of current and voltage, the PP increases with an increase in current and increases with an increase in gap voltage from the lower level to the medium level and then diminishes at the higher level. For the combination of both pulse on time and gap voltage, the PP increases with an increase in pulse on time from a lower level to a higher level, and the removal rate of the material increases with an increase in gap voltage from the lower level to medium level and diminishes at the higher level (Kumar S and Kumar M, 2014). For the combination of current and gap voltage, the PP increases with an increase in current from a lower level to higher level, and PP increases with an increase in voltage from the lower level to medium level and diminishes at the higher level (Kumar *et al.*, 2021).

2.3.5 OPTIMUM PARAMETER MACHINING SURFACE MICRO-STRUCTURAL ANALYSIS

Microstructural analysis was carried out by using a scanning electron microscope on the optimum parameters of the machined surface. Figure 2.12 shows the microstructural examination of the machined surface at the optimal parameters of $IP = 12$ A, $T_{on} = 50$ µs and $V = 40$ V. From the microstructural examination, the formation of microcracks was found on the surface due to lowering of the IP with higher the T_{on}. Formation of microcracks and craters with voids was achieved on the machined surface due to lower the IP with higher the T_{on}. On the machined surface, an increase in T_{on} increases the size of the crater, and microvoids with circularity also enhance and lead to the formation of cavities. At low IP and higher T_{on}, a higher amount of discharge energy was generated, which results in

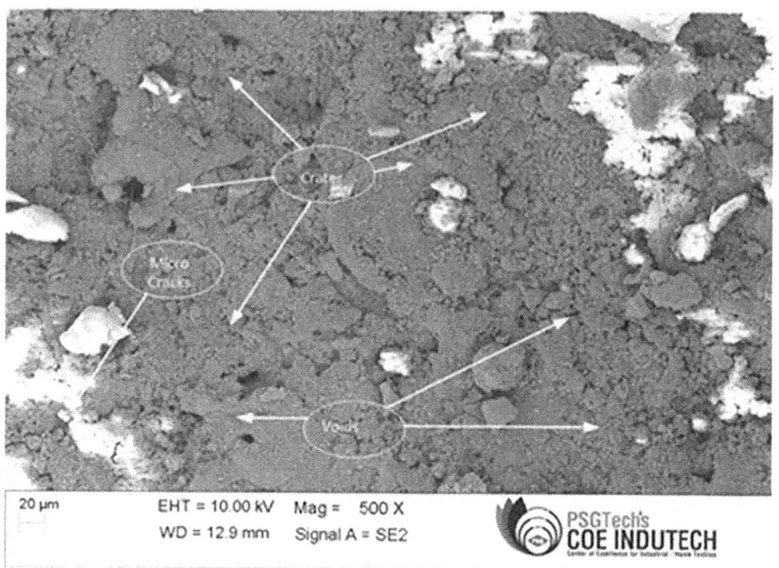

FIGURE 2.12 Machined surface of AA6061/15% SiC composite at a optimum level of $(IP) = 12$ A; $(T_{on}) = 50$ μs; $V = 40$ V.

a higher MRR and forms a cavity on the machined surface, which is due to more plasma being produced between the tool and workpiece. A high V enhances the less amount of precision with a better surface finish obtained. At a lower level, the higher the dissipation energy generated by an IP with a high T_{on} and high V combination, the more metal is removed.

2.3.5.1 Conclusions

AA6061/15% SiC composites were fabricated using stir casting methodology and machined by electrical discharge machining with varying input process parameters to achieve the output performance characteristics. By employing entropy weight methods incorporated with the ARAS method, the output performance characteristics were optimized, and the following conclusions were made:

1. The best optimal process parameters are achieved at the IP of 12 A, T_{on} of 50 μs and V of 40 V to attain higher MRR with lower TWR, CIR, CYL and PP values.
2. From the ARAS utility degree, the optimum conditions achieved at the fourth experimental condition (IP= 12 A; T_{on} = 50 μs; V = 40 V) had a utility degree of 0.8883.
3. The entropy weights for the criteria were 0.087, 0.264, 0.244, 0.147 and 0.249 for MRR, TWR, CIR, CYL and PP.
4. The material removal rate increases with an increase in IP and T_{on}. The tool wear rate decreases with an increase in T_{on} and V. The circularity

increases with an increase in *IP* and T_{on}. The cylindricity decreases with an increase in *IP* and lower T_{on}. The Perpendicularity increases with an increase in *IP* and T_{on}.

5. By using a scanning electron microscope, the machined surface was examined. A better surface finish and microvoids with craters were achieved by increasing the *IP* and T_{on}.

REFERENCES

1 K. H. Ho and S. T. Newman, "State of the Art Electrical Discharge Machining (EDM)," *International Journal of Machine Tools and Manufacture* 43, no. 13 (2003): 1287–1300, https://doi.org/10.1016/S0890-6955(03)00162-7.

2 Tanwir Alam et al., "Mechanical Properties and Morphology of Aluminium Metal Matrix Nanocomposites-Stir Cast Products," *Advances in Materials and Processing Technologies* 3 (2017): 1–16, https://doi.org/10.1080/2374068X.2017.1350543.

3 E. Taheri-Nassaj, M. Kobashi, and T. Choh, "Fabrication of an AlN Particulate Aluminium Matrix Composite by a Melt Stirring Method," *Scripta Metallurgica et Materiala* 32, no. 12 (1995): 1923–1929, https://doi.org/10.1016/0956-716X(95)00083-8.

4 G. Karthik Pandiyan and T. Prabaharan, "Mechanical and Tribological Characterization of Stir Cast AA6061 T6–SiC Composite," *Silicon* 13 (2021): 4575–4582, https://doi.org/10.1007/s12633-020-00781-y.

5 Yuan Feng Chen and Yan Cherng Lin, "Surface Modifications of Al-Zn-Mg Alloy Using Combined EDM with Ultrasonic Machining and Addition of TiC Particles into the Dielectric," *Journal of Materials Processing Technology* 209, no. 9 (2009): 4343–4350, https://doi.org/10.1016/j.jmatprotec.2008.11.013.

6 Velusamy Senthilkumar and Bidwai Uday Omprakash, "Effect of Titanium Carbide Particle Addition in the Aluminium Composite on EDM Process Parameters," *Journal of Manufacturing Processes* 13, no. 1 (2011): 60–66, https://doi.org/10.1016/j.jmapro.2010.10.005.

7 P. Narender Singh, K. Raghukandan, and B. C. Pai, "Optimization by Grey Relational Analysis of EDM Parameters on Machining Al-10%SiCP Composites," *Journal of Materials Processing Technology* 155–156, no. 1–3 (2004): 1658–1661, https://doi.org/10.1016/j.jmatprotec.2004.04.322.

8 Y. W. Seo, D. Kim, and M. Ramulu, "Electrical Discharge Machining of Functionally Graded 15–35 Vol% SiC p/Al Composites," *Materials and Manufacturing Processes* 21, no. 5 (2006): 479–487, https://doi.org/10.1080/10426910500471482.

9 Kavan Panneerselvam and Daniel D. Jafrey, "Study on Tensile Strength, Impact Strength and Analytical Model for Heat Generation in Friction Vibration Joining of Polymeric Nanocomposite Joints," *Polymer Engineering and Science* (2016), https://doi.org/10.1002/pen.24443.

10 T. S. Senthilkumar and R. Muralikannan, "Enhancing the Geometric Tolerance of Aluminium Hybrid Metal Matrix Composite Using EDM Process," *Journal of the Brazilian Society of Mechanical Sciences and Engineering* 41, no. 1 (2019), https://doi.org/10.1007/s40430-018-1553-2.

11 Sweety Mahanta et al., "EDM Investigation of Al 7075 Alloy Reinforced with B4C and Fly Ash Nanoparticles and Parametric Optimization for Sustainable Production,"

Journal of the Brazilian Society of Mechanical Sciences and Engineering 40, no. 5 (2018), https://doi.org/10.1007/s40430-018-1191-8.

12 L. Selvarajan et al., "Modelling and Experimental Investigation of Process Parameters in EDM of Si_3N_4-TiN Composites Using GRA-RSM," *Journal of Mechanical Science and Technology* 31, no. 1 (2017): 111–122, https://doi.org/10.1007/s12206-016-1009-5.

13 L. Selvarajan et al., "Optimization of EDM Process Parameters in Machining Si_3N_4-TiN Conductive Ceramic Composites to Improve Form and Orientation Tolerances," *Measurement: Journal of the International Measurement Confederation* 92 (2016): 114–129, https://doi.org/10.1016/j.measurement.2016.05.018.

14 Jurgita Antucheviciene, Mehdi Keshavarz-Ghorabaee, Edmundas Kazimieras Zavadskas, and Zenonas Turskis, "A New Combinative Distance-Based Assessment," *Economic Computation and Economic Cybernetics Studies and Research* 50, no. 1 (2016): 39–68.

15 L. Selvarajan, C. Sathiya Narayanan, and R. Jeyapaul, "Optimisation of EDM Parameters on Machining Si_3N_4-TiN Composite for Improving Circularity, Cylindricity and Perpendicularity," *Materials and Manufacturing Processes* (August 2016), https://doi.org/10.1080/10426914.2015.1058947.

16 Deepti Ranjan Sahu and Amitava Mandal, "Critical Analysis of Surface Integrity Parameters and Dimensional Accuracy in Powder-Mixed EDM," *Materials and Manufacturing Processes* (2020): 1–12, https://doi.org/10.1080/10426914.2020.1718695.

17 Vineet Kumar Yadav, Pradeep Kumar, and Akshay Dvivedi, "Effect of Tool Rotation in Near-Dry EDM Process on Machining Characteristics of HSS," *Materials and Manufacturing Processes* (2019): 1–12, https://doi.org/10.1080/10426914.2019.1605171.

18 Prosun Mandal and Subhas Chandra Mondal, "Surface Characteristics of Mild Steel Using EDM with Cu-MWCNT Composite Electrode," *Materials and Manufacturing Processes* (2019): 1–7, https://doi.org/10.1080/10426914.2019.1605179.

19 Farshid Jafarian, "Electro Discharge Machining of Inconel 718 Alloy and Process Optimization," *Materials and Manufacturing Processes* (2020): 1–9, https://doi.org/10.1080/10426914.2020.1711919.

20 Y. L. Teng et al., "Machining Characteristics of PCD by EDM with Cu-Ni Composite Electrode," *Materials and Manufacturing Processes* 6914 (2020), https://doi.org/10.1080/10426914.2020.1718700.

21 Ankita Sarmah et al., "Surface Modification of Aluminum with Green Compact Powder Metallurgy Inconel-Aluminum Tool in EDM," *Materials and Manufacturing Processes* (2020): 1–9, https://doi.org/10.1080/10426914.2020.1765253.

22 Sudhir Kumar, Sanjoy Kumar Ghoshal, and Pawan Kumar Arora, "Multi-Variable Optimization in Die-Sinking EDM Process of AISI420 Stainless Steel," *Materials and Manufacturing Processes* (2020): 1–11, https://doi.org/10.1080/10426914.2020.1843678.

23 N. Manikandan et al., "Machinability Analysis and ANFIS Modelling on Advanced Machining of Hybrid Metal Matrix Composites for Aerospace Applications," *Materials and Manufacturing Processes* 34, no. 16 (2019): 1866–1881, https://doi.org/10.1080/10426914.2019.1689264.

24 Naveed Ahmed et al., "Machinability of Titanium Alloy Through Electric Discharge Machining," *Materials and Manufacturing Processes* 34, no. 1 (2019): 93–102, https://doi.org/10.1080/10426914.2018.1532092.

25 Van Tao Le, "The Influence of Additive Powder on Machinability and Surface Integrity of SKD61 Steel by EDM Process," *Materials and Manufacturing Processes* (2021), 1–15, https://doi.org/10.1080/10426914.2021.1885710.

26 Md Al-Amin et al., "Multiple-Objective Optimization of Hydroxyapatite-Added EDM Technique for Processing of 316L-Steel," *Materials and Manufacturing Processes* (2021): 1–12, https://doi.org/10.1080/10426914.2021.1885715.

27 Jitendra Kumar et al., "Machining and Optimization of Zircaloy-2 Using Different Tool Electrodes," *Materials and Manufacturing Processes* (2021): 1–11, https://doi.org/10.1080/10426914.2021.1905829.

28 S. Vinoth Kumar and M. Pradeep Kumar, "Optimization of Cryogenic Cooled EDM Process Parameters Using Grey Relational Analysis," *Journal of Mechanical Science and Technology* 28, no. 9 (2014): 3777–3784, https://doi.org/10.1007/s12206-014-0840-9.

29 S. Vinoth Kumar and M. Pradeep Kumar, "Machining Process Parameter and Surface Integrity in Conventional EDM and Cryogenic EDM of Al-SiCp MMC," *Journal of Manufacturing Processes* 20 (2015): 70–78, https://doi.org/10.1016/j.jmapro.2015.07.007.

30 S. Vinoth Kumar and M. Pradeep Kumar, "Experimental Investigation of the Process Parameters in Cryogenic Cooled Electrode in EDM," *Journal of Mechanical Science and Technology* 29, no. 9 (2015): 3865–3871, https://doi.org/10.1007/s12206-015-0832-4.

31 G. Karthik Pandiyan et al., "Improvement on the Machinability Characteristics of AA6061-T6/15 Wt.% SiC Composites by Response Surface Methodology," *Surface Topography: Metrology and Properties* 9, no. 3 (2021): 35050, https://doi.org/10.1088/2051-672X/ac2560.

32 Edmundas Kazimieras Zavadskas and Zenonas Turskis, "A New Additive Ratio Assessment (ARAS) Method in Multicriteria Decision-Making," *Technological and Economic Development of Economy* 16, no. 2 (2010): 159–172, https://doi.org/10.3846/tede.2010.10.

33 Edmundas Kazimieras Zavadskas et al., "Selection of the Effective Dwelling House Walls by Applying Attributes Values Determined at Intervals," *Journal of Civil Engineering and Management* 14, no. 2 (2008): 85–93, https://doi.org/10.3846/1392-3730.2008.14.3.

34 Alireza Alinezhad and Javad Khalili, "ARAS Method BT." *New Methods and Applications in Multiple Attribute Decision Making (MADM)*, eds. Alireza Alinezhad and Javad Khalili (Cham: Springer International Publishing, 2019), 67–71, https://doi.org/10.1007/978-3-030-15009-9_9.

35 M. Ghram and H. Frikha, "Multiple Criteria Hierarchy Process within ARAS Method," in *2019 6th International Conference on Control, Decision and Information Technologies (CoDIT)*, 2019, 995–1000, https://doi.org/10.1109/CoDIT.2019.8820401.

36 Karthik Pandiyan and Prabaharan, "Mechanical and Tribological Characterization of Stir Cast AA6061 T6–SiC Composite." *Silicon* 13 (2021): 4575–4582, https://doi.org/10.1007/s12633-020-00781-y.

Enhancing Electro Discharge Machining Performance for LM25 Aluminium Composite Using Mathematical Modelling and GRA-Guided Particle Swarm Optimization Technique

R. Rajesh, M. Dev Anand,
K.S. Jai Aultrin, and S. Raja

NOMENCLATURE

EDM	Electric discharge machining
V_d	Discharge voltage
I_p	Discharge current
T_{off}	Pulse-off time
T_{on}	Pulse-on time
P_{oil}	Oil pressure
S_g	Spark gap
MRR	Material removal rate
SR	Surface roughness
GRA	Gray relational analysis
PSO	Particle swarm optimization
ANOVA	Analysis of variance
RSM	Response surface methodology
GRC	Grey relational coefficient
S/N	Signal-to-noise
Γ	Predicted response
Γm	Mean value of response
Γj	Mean value of response at the optimal level

DOI: 10.1201/9781003397465-3

3.1 INTRODUCTION

EDM is a widely recognized method predominantly employed for precision machining of intricate contoured work pieces, serving as an alternative to conventional procedures. This advanced technique is not only instrumental in achieving highly accurate results but also provides valuable insights into the physical phenomena intrinsic to the process. To ensure consistency and efficiency in EDM applications, it is imperative to standardize the methodology. Rapid advancements in manufacturing technology have increased the importance of non-traditional machining (NTM) procedures not only in contemporary machining but also in the cost-effective machining of materials that are generally too complex to machine with traditional tools [1–3, 21]. Predictions derived from the semi-empirical model, with model parameters calculated using a non-linear optimization method, have demonstrated good agreement with experimental results [4]. EDM has been explored by a multidisciplinary team of researchers. They optimized process parameters for distinct kinds of EDM at various points in time using disparate optimization models and solution methodologies. To attain a greater MRR in EDM, a stable machining process is required, where contamination impacts the distance between the work piece and the tool, and the eroding surface size also affects the machining command set on the tool [5, 6]. MRR and TWR on AISIH13 tool steel were examined. I_p was discovered to be the primary component influencing MRR. With a high I_p, a medium T_{on}, and a low T_{off}, a higher MRR was attained. However, a lower TWR was attained with a high I_p, a high T_{on}, and a low T_{off} value [7]. MRR on cobalt-cemented carbide/ tungsten carbide was studied using electrode rotation, T_{on}, I_p, and dielectric flushing pressure (FP), and it was shown in an experimental setup that I_p and T_{on} are the most important parameters. Kuppan et al. [8] proposed a mathematical model for the MRR of Inconel 718 in a deep hole drilling process. The experiment was designed to simulate the same phenomenon using CCD and response surface methodology (RSM). The peak current and duty factor significantly influence the material removal rate (MRR), and by utilizing the desire function approach, the machining parameters have been optimized to achieve a higher MRR. Puertas et al. [9] investigated the influence of EDM settings on electrode wear and MRR in a cobalt-bonded tungsten carbide work piece. MRR was calculated for each and every response using a quadratic model; the most relevant element was current intensity, followed by T_{on}, T_{off}, and the interactive effect between the first two. MRR was maximized first, and subsequently current intensity and T_{off} were maximized and reduced using T_{on}. Khan [10] addressed the performance (TWR and MRR) of EDM mildsteel in relation to copper and brass electrodes. Round electrodes had the highest MRR, followed by triangular, square, and diamond-shaped electrodes. However, the diamond-shaped electrodes had the greatest EWR. Diamond-shaped electrodes is also regarded as a strategy for offline process planning. The method used in the simulation was highly dependent on MRR and TWR. The position of the discharge and the reproduction of the spark gap depend on the concentration of contaminants in the dispersion, which was shown to generate a highly practical depiction of sparking occurrences.

Khan et al. [11] followed up with a study of the overall performance of brass and copper electrodes, noting that the highest MRR was recorded while machining aluminium with brass electrodes. The electrode material, such as brass, had poor heat conductivity, and practically all of the heat energy was utilized to remove unwanted material from the aluminium work piece, which had a low melting point. Dhar et al. [12] calculated the influence of T_{on}, I_p, and V_d on the TWR, MRR, and S_g on the EDM of Al-4Cu-6Si alloy10wt.%SiCP composites. A nonlinear mathematical model of the second order was developed to illustrate the relationship between different factors in machining. It was noticed that when T_{on} and I_p grew, S_g, TWR, and MRR increased as well. Salonitis et al. [13] constructed a simple temperature-based model for calculating the MRR and concluded that increasing I_p, T_{on}, or V_{on} leads to a bigger MRR; also, MRR rises when T_{off} decreases. They concluded that model predictions and experimental data accord well. E.L. Taweel [14] investigated the process parameters in EDM of Al-Cu-Si-TiC and CK45 steel composites formed by the powder metallurgy process and assessed TWR and MRR. It has been shown that these electrodes are highly sensitive to T_{on} and I_p compared to ordinary conventional electrodes. To achieve a high MRR and a low TWR, the design was adjusted, and the results were proven experimentally. Chiang [15] established the influence of V_{on}, T_{on}, I_p, and T_{off} on EWR and MRR reactions. The studies were organized using a central composite design on an Al_2O_3+TiC work piece, and the effects of factors were investigated using ANOVA communications.

A sophisticated mathematical model was proposed and claimed to accurately predict and fit MRR with a confidence level of 95%. T_{off} and I_p were the most essential characteristics that influenced the response. Akhay et al. [16] determined EDM machining performance in terms of TWR and MRR by optimising the T_{on}, T_{off}, I_p, and FP settings in an aluminium 6063 SiCp metal matrix composite (MMC). When compared to other significant factors, it was shown that I_p was predominant on MRR. MRR value increases in lockstep with T_{on} and I_p until it reaches an ideal level, at which time it is halted. Karthikeyan et al. [17] used full factorial design to develop a mathematical model for optimizing EDM parameters like as TWR, MRR, and SR on aluminium SiCp composites (FFD). The percentage volume fractions of SiC, V_d, T_{on}, and I_p contained in LM25 aluminium MMCs were considered. Wang [18] investigated the optimization and practicality of EDM for testing the machinability of composite materials such as W/Cu by employing the L18 orthogonal array table to determine the polarity, T_{on}, I_p, T_{off}, rotating electrode planetorial speed, and Vd in order to determine the TWR and MRR. TWR is comparable to the MRM in EDM. Mohri et al. [19] established that precipitation occurs during sparking and has an effect on tool wear as a result of turbo static carbon precipitation of hydrocarbon dielectrics on the electrode surface. Additionally, a shortage of carbon precipitating in a complicated manner led to the abrupt wear on the electrode's edge. Marafona and Wykes [20] determined the composition of tool surfaces using energy dispersive X-ray analysis. When fine-tuning the parameter settings for standard EDM circumstances, a wear stopper or inhibitor carbon layer was placed on the electrode surface. Due to carbon thickness in inhibitor layer, there is a noticeable increase in TWR;

nevertheless, it has a negligible influence on the MRR. On the other hand, when a greater MRR is required, a high pulse current is encouraged to enhance electrode wear by implanting electrode material into the work piece. Mohri et al. [19] and Bleys et al. [21] developed a system for online tool wear compensation based on pulse analysis and managed the tool's dynamic feed movement. Kunieda and Kobayashi [22] analysed the TWR ratio using spectroscopic analysis of the tool electrode's vapour density. It is well established that a longer T_{on} results in a lower TWR and the deposition of a thicker coating of carbon on the tool electrode surface. Conversely, when the carbon layer was thicker, the density of copper vapour evaporated off the surface of the tool electrode was found to be lower, showing that the carbon layer's protective properties prevented tool electrode wear.

Sameh et al. [23] demonstrated the advancement of a broad numerical model in associating the higher-order manipulation techniques and interactivity of several EDM parameters that are transferred by RSM, applying relevant experimental values gathered during testing. The mathematical models were constructed using RSM, which utilizes data values obtained from real observations of work piece EDM. Exploration was conducted to determine the necessary control parameters for the MRR, EWR, gap size, and Ra.

Staelens and Kruth [24] and Yu et al. [25] described a technique for compensating longitudinal TWR by extending the forward and backward machining movement. Mahdavi Nejad [26] reported work aiming at simultaneously optimizing the SR and MRR of EDM of SiC parameters. Because the output parameters are incompatible, there is no single combination of machining parameters that provides optimum machining performance. Intelligent algorithms were used in conjunction to process the model. To optimize the process parameters, the non-dominating sorting genetic algorithm II (NSGA-II) and the MOO technique were utilized. Some key repercussions of the input process parameters were considered: the I_p, T_{off}, T_{on}, and EDM of SiC. Experiments were done across a wide range of input process parameters considered for the model's training and verification. Yilmaz et al. [27] built a user-friendly fuzzy expert system for selecting EDM settings. Investigations substantiated the system and fuzzy model. Defuzzification techniques, fuzzy expert rules (if–then rules), and membership functions all contributed to resolving the tough problem. As a result of the conclusion, a supplemental fine selection of EDM complicated parameters to compute was input using the accepted fuzzy model. Ritam Choudhary et al. [28] focused on optimizing the process parameters using the Taguchi grey analysis (TGA) method for the wired electro discharge machining (WEDM) process of tungsten. Çakıroğlu [29] developed mathematical models using artificial neural network (ANN) and regression analysis (RA) methods for predicting output parameters. When the two methods were compared, it was seen that the ANN model is closer to the experiment.

This research investigates each performance metric and devises a strategy aimed at enhancing MRR and surface roughness (SR). Previous efforts have significantly enhanced the productivity, precision, safety, and adaptability of the EDM process. The existing literature indicates a focus on a maximum of four input potential parameters for the optimization of EDM. However, in this study, six input parameters were meticulously chosen based on the machine's inherent capabilities. This selection was made considering the available facility within the machine, aiming to maximize the potential for optimization and efficiency. The main aim is to select six key processing parameters—current (I_p), voltage (V_d), pulse-on (T_{on}), pulse-off (T_{off}), spark gap (S_g), and oil pressure (P_{oil})—with a dual focus. These parameters are selected not only to enhance MRR but also to improve accuracy while simultaneously ensuring SR.

3.2 EXPERIMENTAL INVESTIGATION

3.2.1 WORK PIECE SELECTION AND DESIGN PARAMETERS

In this experiment, a composite plate of LM25 (Al-SiCp [10%]) measuring 120 × 120 × 8 mm is manufactured using the stir casting method, intended for applications in aerospace and the manufacturing of wheels in the transport industry. The composition of the material, physical, and mechanical characteristics are reported in Tables 3.1 and 3.2. The studies were done under a variety of machining settings using a 3HP/2.2kW Electronica 5030 Die Sinking EDM machine. The input process parameter was selected based on the geometry of the work piece surfaces and the setting machining. The tests were conducted according to the standard

TABLE 3.1

Composition of LM25 (Al-SiCp [10%]) Composite

Work Material	LM25 (Al-SiCp [10%])
Mg (%)	0.45
Si (%)	7.5
Cu (%)	0.2
Mn (%)	0.1
Fe (%)	0.2
Zn (%)	0.1
Ti (%)	0.2
SiC (%)	10
Reinforcement	10% SiC particles (by volume)
Particle size (μm)	20

TABLE 3.2

Physical and Mechanical Properties of LM25 (Al-SiCp [10%]) Composite

Material	Density (gms/cm³)	Tensile Strength (N/mm²)	Hardness (BHN)	Modulus of Elasticity (x103 N/mm²)	Percentage of Elongation (%)
LM25 (Al-SiCp [10%])	2.68	275	110	90	1.2–1.8

technique outlined previously. Figure 3.1a shows a close up of the plate blank used to cut the specimens put on the EDM machine, while Figure 3.1c shows the machined work piece. As seen in Table 3.3, the levels were determined specifically for each process parameter. Six process factors at three different levels resulted in a total of 54 machining operation tests. Following each test, the work piece is measured using the SR tester SJ201 (Fig.3.1d) with a 5-m tip radius to determine the SR, and the MRR is computed using equation 3.1

$$\text{MRR (mg/Min)} = \frac{\left(\text{Initial weight} - \text{Final weight}\right)}{\text{Machining time}} \tag{3.1}$$

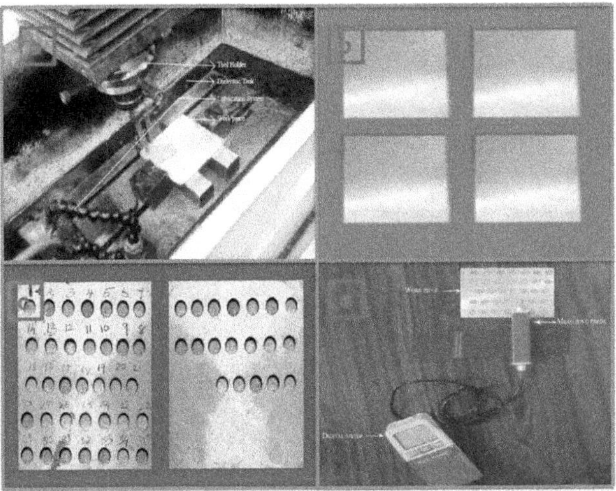

FIGURE 3.1   a) Plate material blank mounted on EDM machine; b) blank material; c) machined work piece specimens; d) surface roughness tester.

TABLE 3.3

Variables Used in the Experiment and Their Levels of LM25 Al Composite

Variable	Coding	Level 1	Level 2	Level 3
Discharge voltage (V_d) in V	A	60	65	70
Discharge current (I_p) in A	B	5	10	15
Pulse-on time (T_{on}) in s	C	15	30	45
Pulse-off time (T_{off}) in s	D	5	7	9
Spark gap (S_g) in mm	E	0.1	0.2	0.3
Oil pressure (P_{oil}) in kg/cm^2	F	1	1.5	3

3.2.2 Design of Experiment

The tests are conducted using RSM using a Box-Behnken technique with six variables at their respective levels. The average number of tests performed in conjunction with the EDM machining parameter for LM25 Al composite is shown in Table 3.4, and the corresponding measured MRR and SR are shown in Table 3.13.

TABLE 3.4

Planning Matrix of the Experiments with the Optimal Model Data for LM25 Al Composite

Sl. No.	A Voltage (V)	B Current (A)	C Pulse-on Time (s)	D Pulse-off Time (s)	E Gap (mm)	F Oil Pressure (Kg/cm²)
1.	65	5	15	7	0.3	1.5
2.	75	10	45	7	0.2	2.0
3.	75	10	30	9	0.1	1.5
4.	65	15	45	7	0.3	1.5
5.	75	15	30	5	0.2	1.5
6.	75	5	30	5	0.2	1.5
7.	65	10	15	9	0.2	1.0
8.	75	15	30	9	0.2	1.5
9.	75	10	45	7	0.2	1.0
10.	65	5	45	7	0.3	1.5
11.	60	10	30	9	0.1	1.5
12.	60	5	30	5	0.2	1.5
13.	60	10	30	5	0.3	1.5
14.	60	10	30	9	0.3	1.5
15.	65	5	30	7	0.1	1.0
16.	65	10	30	7	0.2	1.5
17.	60	10	45	7	0.2	2.0
18.	65	5	30	7	0.3	2.0
19.	65	10	15	5	0.2	2.0
20.	60	5	30	9	0.2	1.5
21.	75	10	30	5	0.1	1.5
22.	65	15	15	7	0.1	1.5
23.	75	5	30	9	0.2	1.5
24.	75	10	30	5	0.3	1.5
25.	75	10	15	7	0.2	2.0
26.	65	10	15	9	0.2	2.0
27.	65	5	30	7	0.1	2.0
28.	65	10	45	5	0.2	2.0
29.	65	5	30	7	0.3	1.0
30.	65	15	30	7	0.1	1.0

(Continued)

TABLE 3.4 *(Continued)*

Sl. No.	A Voltage (V)	B Current (A)	C Pulse-on Time (s)	D Pulse-off Time (s)	E Gap (mm)	F Oil Pressure (Kg/cm²)
31.	65	15	30	7	0.3	2.0
32.	65	10	30	7	0.2	1.5
33.	65	10	45	9	0.2	1.0
34.	60	10	15	7	0.2	1.0
35.	65	10	45	9	0.2	2.0
36.	65	10	45	5	0.2	1.0
37.	65	5	15	7	0.1	1.5
38.	75	10	15	7	0.2	1.0
39.	65	15	30	7	0.1	2.0
40.	65	15	15	7	0.3	1.5
41.	65	10	30	7	0.2	1.5
42.	75	10	30	9	0.3	1.5
43.	65	10	30	7	0.2	1.5
44.	65	10	30	7	0.2	1.5
45.	65	15	45	7	0.1	1.5
46.	60	15	30	5	0.2	1.5
47.	60	10	45	7	0.2	1.0
48.	65	15	30	7	0.3	1.0
49.	60	15	30	9	0.2	1.5
50.	65	10	30	7	0.2	1.5
51.	65	10	15	5	0.2	1.0
52.	60	10	15	7	0.2	2.0
53.	65	5	45	7	0.1	1.5
54.	60	10	30	5	0.1	1.5

3.2.3 MULTIPLE REGRESSION ANALYSIS

The experimental data from EDM drilling are examined to generate a multi-regression equation. Along with the experimental data, the regression equation is utilized to create sufficient data to train the suggested prediction networks. MRR and SR are dependent on I_p, V_d, S_g, T_{on}, T_{off}, and P_{oil}, which is created using multi-regression equations 3.2 and 3.3.

$$\begin{aligned}
\mathbf{MRR} = {}& 1.6059 - 0.2872B + 0.0852A - 0.0740C - 4.1745E \\
& - 0.1632D - 0.9694F - 0.0008A2 - 0.00C2 - 0.0071B2 \\
& + 0.05D2 - 7.6986E2 + 0.0084AB - 0.4526F2 + 0.0006AC \\
& - 0.0056AD - 0.0045AF - 0.0663AE + 0.0125BC \\
& + 0.1064BF - 0.0082BD - 0.5275BE + 0.1433CE \\
& - 0.0197CF + 0.0008CD - 0.6375DE + 0.0219DF + 10.1EF
\end{aligned} \tag{3.2}$$

$$
\begin{aligned}
\mathbf{SR} \; = \; & 4.805 - 0.1893A + 0.6101C + 1.2479B - 0.4769D - 3.2685F \\
& - 41.5598E - 0.0163B2 + 0.0021A2 - 0.0023C2 + 22.4167E2 \\
& + 0.0165D2 + 1.3300F2 - 0.0116AB + 0.0060AD - 0.0043AC \\
& + 0.4410AE - 0.0007AF - 0.0249BD + 0.0038BC + 0.0295BF \\
& - 0.0172CD + 0.1575BE + 0.1058CE - 0.0368CF + 0.2025DF \\
& + 1.1625DE - 6.9250EF
\end{aligned}
\tag{3.3}
$$

3.2.4 ANOVA-Based Prediction Using Mathematical Modelling

The machining parameters' (I_d, V_d, S_g, T_{off}, T_{on} and P_{oil}) influence on the variables MRR and SR for composites was examined via tests, as described in Table 3.5. In finding the association among the input process parameters and the output responses of MRR and SR, Minitab software is employed. The values of the coefficient of determination (R^2) and the adjusted R^2 statistic (R^2adj) values are compared, as described in Table 3.6 for MRR and Table 3.8 for SR, to determine the regression model. The full quadratic model for MRR and SR is the best among all models before backward elimination, as listed in Tables 3.5 and 3.7, where R^2 = 98.02% for MRR and R^2 = 99.18% for SR indicate that the 98.02% and 99.18% differences in responses are described by analysts in the model. However, the R^2adj values are 95.96% for MRR and 98.32% for SR, indicating that the model incorporates a significant amount of analysis that highlights the relevance of the relationships. Therefore, for deeper investigation, study of the entire quadratic model is adopted. Tables 3.5 and 3.7 depict the regression coefficients in coded units and their significance in the model. The columns in the table correspond to the terms, the coefficients (Coeff.), the standard error of the coefficients (SE Coeff.), the t-statistic, and the p-value, which are used in turn to determine whether to reject or fail to reject the null hypothesis. To assess the appropriateness of the model with a confidence level of 95%, the p-value of the statistically significant term should be less than 0.05.

The values marked * in the last column of the table exceed the 0.05 value. Hence, these concepts are irrelevant and eliminated for the next study. Thus the backward elimination process omits the irrelevant terms to control the built-in quadratic model. After the exclusion, the model is provided in Figure 3.3 and Table 3.6 for MRR and Table 3.8 for SR. After backward elimination, the values of R^2 and R^2adj are 98.27% and 97.65% for MRR and 99.11% and 98.32% for SR, respectively.

The reduced model has a higher R^2 than the full quadratic model (98.27%), with adjusted R^2 values of 97.65% for MRR and 99.20% and 98.32% for surface roughness (SR). This indicates significant relationships among the reaction variables and conditions. After eliminating terms from the acceptable model, the remaining factors are A*A, B*B, C*C, E*E, F*F, A*C, A*D, A*E, A*F, B*D, C*D, D*E, and D*F for MRR, as well as D*D and A*F for SR.

TABLE 3.5
ANOVA Table for MRR Estimated Regression Coefficients (Before Backward Elimination)

Source	DF	Adj SS	Adj MS	F-Value	P-Value
Model	27	211.070	7.817	47.62	0.000
Linear	6	181.381	30.230	184.15	0.000
A	1	0.570	0.570	3.47	0.074*
B	1	136.049	136.049	828.77	0.000
C	1	43.897	43.897	267.41	0.000
D	1	0.004	0.004	0.02	0.883*
E	1	0.860	0.860	5.24	0.030
F	1	0.001	0.001	0.00	0.948*
Square	6	1.445	0.241	1.47	0.228*
A*A	1	0.016	0.016	0.09	0.761*
B*B	1	0.328	0.328	2.00	0.169*
C*C	1	0.000	0.000	0.00	0.988*
D*D	1	0.412	0.412	2.51	0.125*
E*E	1	0.061	0.061	0.37	0.548*
F*F	1	0.132	0.132	0.80	0.379*
2-way interaction	15	12.665	0.844	5.14	0.000
A*B	1	0.853	0.853	5.20	0.031
A*C	1	0.040	0.040	0.25	0.624*
A*D	1	0.118	0.118	0.72	0.404*
A*E	1	0.021	0.021	0.13	0.722*
A*F	1	0.002	0.002	0.02	0.903*
B*C	1	6.998	6.998	42.63	0.000
B*D	1	0.054	0.054	0.33	0.573*
B*E	1	1.113	1.113	6.78	0.015
B*F	1	0.566	0.566	3.45	0.075*
C*D	1	0.005	0.005	0.03	0.863*
C*E	1	0.370	0.370	2.25	0.145*
C*F	1	0.351	0.351	2.14	0.156*
D*E	1	0.130	0.130	0.79	0.382*
D*F	1	0.004	0.004	0.02	0.880*
E*F	1	2.040	2.040	12.43	0.002
Error	26	4.268	0.164		
Lack-of-fit	21	3.264	0.155	0.77	0.695*
Pure error	5	1.004	0.201		
Total	53	215.338			

Model Summary

S	R-sq	R-sq (Adj)
0.405165	98.02%	95.96%

TABLE 3.6
ANOVA Table for MRR Estimated Regression Coefficients (After Backward Elimination)

Source	DF	Adj SS	Adj MS	F-Value	P-Value
Model	14	210.970	15.069	158.59	0.000
Linear	6	186.596	31.099	327.28	0.000
A	1	0.785	0.785	8.26	0.007
B	1	133.963	133.963	1409.79	0.000
C	1	50.753	50.753	534.11	0.000
D	1	0.011	0.011	0.12	0.002
E	1	1.078	1.078	11.34	0.002
F	1	0.006	0.006	0.06	0.006
Square	1	1.089	1.089	11.46	0.002
D*D	1	1.089	1.089	11.46	0.002
2-way interaction	7	13.133	1.876	19.74	0.000
A*B	1	0.889	0.889	9.35	0.004
B*C	1	6.998	6.998	73.64	0.000
B*E	1	0.863	0.863	9.08	0.005
B*F	1	0.866	0.866	9.11	0.004
C*E	1	0.370	0.370	3.89	0.046*
C*F	1	0.567	0.567	5.96	0.019
E*F	1	2.581	2.581	27.16	0.000
Error	39	3.706	0.095		
Lack-of-fit	34	3.134	0.092	0.81	0.289
Pure error	5	0.572	0.114		
Total	53	214.676			

Model Summary

S	R-sq	R-sq (adj)
0.308259	98.27%	97.65%

In judging the adequacy of the second-order model, ANOVA is used, including a significance test, coefficients of the model, and lack-of-fit test. Tables 3.7 and 3.10 provide information on MRR and SR, while Table 3.9 summarizes the ANOVA for the given model. This model incorporates two sources of variation, regression and residual error. The tables offer insights into the effects of these sources on the overall machining process, aiding in the understanding and optimization of the studied parameters. The variation due to the terms in the model is the summation of linear and square terms, whereas lack of fit and pure error contribute to residual error. Tables 3.7 and 3.9 depict the sources of variation, adjusted sum square error (Adj SS), degrees of freedom (DoF), sequential sum square error (SeqSS), adjusted mean square error (Adj MS), F-statistic and the p-values in columns. The p-value for lack of fit is less than 0.05, indicating that the results are statistically significant at the 95% confidence level.

TABLE 3.7
Analysis of Variance for MRR

Source	DF	Seq SS	Adj SS	Adj MS	F	P
Regression	27	211.070	211.070	7.8174	47.62	0.000
Linear	6	196.961	181.381	30.2302	184.15	0.000
Square	6	1.445	1.445	0.2408	1.47	0.000
Interaction	15	12.665	12.665	0.8443	5.14	0.000
Residual error	26	4.268	4.268	0.1642		
Lack-of-fit	21	3.264	3.264	0.1554	0.77	0.295
Pure error	5	1.004	1.004	0.2008		
Total	53	215.338				

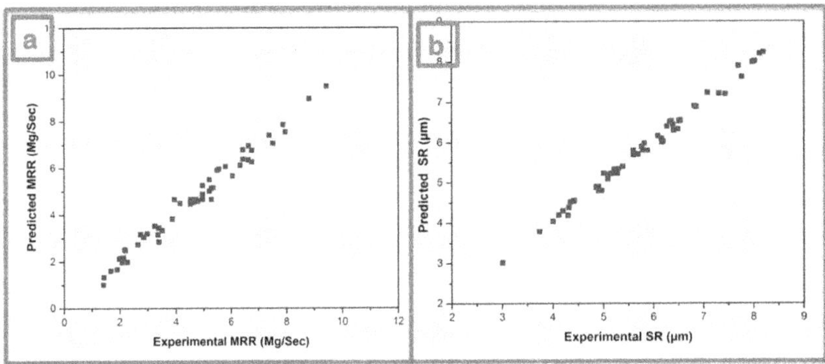

FIGURE 3.2   Comparison plot between experimental and predicted: a) MRR; b) SR.

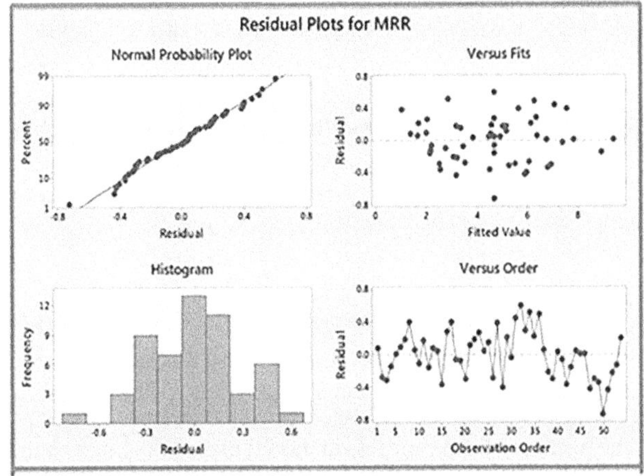

FIGURE 3.3   Residual plot for MRR.

TABLE 3.8

ANOVA Table for SR Estimated Regression Coefficients (Before Backward Elimination)

Source	DF	Adj SS	Adj MS	F-Value	P-Value
Model	27	63.3113	2.3449	116.02	0.000
Linear	6	32.6035	5.4339	268.87	0.000
A	1	0.7315	0.7315	36.20	0.000
B	1	11.5816	11.5816	573.06	0.000
C	1	19.8213	19.8213	980.77	0.000
D	1	0.4655	0.4655	23.03	0.000
E	1	0.0009	0.0009	0.04	0.835*
F	1	0.0027	0.0027	0.14	0.716*
Square	6	5.3204	0.8867	43.88	0.000
A*A	1	0.0968	0.0968	4.79	0.038
B*B	1	2.1151	2.1151	104.66	0.000
C*C	1	2.4752	2.4752	122.47	0.000
D*D	1	0.0386	0.0386	1.91	0.179*
E*E	1	0.3251	0.3251	16.09	0.000
F*F	1	0.6146	0.6146	30.41	0.000
2-way interaction	15	12.3686	0.8246	40.80	0.000
A*B	1	1.5242	1.5242	75.42	0.000
A*C	1	1.9837	1.9837	98.15	0.000
A*D	1	0.1333	0.1333	6.59	0.016
A*E	1	0.8684	0.8684	42.97	0.000
A*F	1	0.0020	0.0020	0.10	0.754*
B*C	1	0.6441	0.6441	31.87	0.000
B*D	1	0.4955	0.4955	24.52	0.000
B*E	1	0.3192	0.3192	15.80	0.000
B*F	1	0.3160	0.3160	15.64	0.001
C*D	1	2.1208	2.1208	104.94	0.000
C*E	1	0.2016	0.2016	9.98	0.004
C*F	1	1.2227	1.2227	60.50	0.000
D*E	1	0.4320	0.4320	21.37	0.000
D*F	1	0.3285	0.3285	16.25	0.000
E*F	1	1.7766	1.7766	87.91	0.000
Error	26	0.5255	0.0202		
Lack-of-fit	21	0.4331	0.0206	1.12	0.498*
Pure error	5	0.0924	0.0185		
Total	53	63.8367			

Model summary S = 0.142162, R-sq = 99.18%, R-sq (adj) = 98.32%

However, the p-value of the regression model and all its linear and square terms have p-value 0.000; ergo, they are statistically significant at 95% confidence, and so the model appropriately describes the experimental data. Multiregression analysis is undertaken to create a quadratic response surface model for MRR and SR,

TABLE 3.9

ANOVA Table for SR Estimated Regression Coefficients (After Backward Elimination)

Source	DF	Adj SS	Adj MS	F-Value	P-Value
Model	25	63.2707	2.5308	125.19	0.000
Linear	6	32.6023	5.4337	268.78	0.000
A	1	0.7315	0.7315	36.18	0.000
B	1	11.5816	11.5816	572.88	0.000
C	1	19.8213	19.8213	980.45	0.000
D	1	0.4655	0.4655	23.02	0.000
E	1	0.0009	0.0009	0.04	0.035
F	1	0.0015	0.0015	0.08	0.048
Square	5	5.2819	1.0564	52.25	0.000
A*A	1	0.1453	0.1453	7.19	0.012
B*B	1	2.2433	2.2433	110.96	0.000
C*C	1	2.6177	2.6177	129.48	0.000
E*E	1	0.3000	0.3000	14.84	0.001
F*F	1	0.5833	0.5833	28.85	0.000
2-way Interaction	14	12.3666	0.8833	43.69	0.000
A*B	1	1.5242	1.5242	75.40	0.000
A*C	1	1.9837	1.9837	98.12	0.000
A*D	1	0.1333	0.1333	6.59	0.016
A*E	1	0.8684	0.8684	42.96	0.000
B*C	1	0.6441	0.6441	31.86	0.000
B*D	1	0.4955	0.4955	24.51	0.000
B*E	1	0.3192	0.3192	15.79	0.000
B*F	1	0.3160	0.3160	15.63	0.000
C*D	1	2.1208	2.1208	104.90	0.000
C*E	1	0.2016	0.2016	9.97	0.004
C*F	1	1.2227	1.2227	60.48	0.000
D*E	1	0.4320	0.4320	21.37	0.000
D*F	1	0.3285	0.3285	16.25	0.000
E*F	1	1.7766	1.7766	87.88	0.000
Error	28	0.5661	0.0202		
Lack-of-fit	23	0.4737	0.0206	1.11	0.401
Pure error	5	0.0924	0.0185		
Total	53	63.8367			

Model summary S = 0.142185, R-sq = 99.20%, R-sq (adj) = 98.32%

and the equation thus derived in uncoded units is presented in Equations 3.3 and 3.4. In examining the ramifications of machining settings on the responses of MRR and SR on EDM in the LM25 Al composite, a mathematical model may be applied. Expected responses are computed using the fitted model and the residuals from the discrepancies between the fitted and observed responses. Table 3.13

displays the machining process parameters for each run order, along with experimental findings (Expt.), the predicted response (Pred.), and the residues (Resi.), where the residues are the difference between the experimentally observed data and the model predictions. The predicted values of MRR and SR generated using Equations 3.3 and 3.4 are close to the experimental values, confirming the model's adequacy (Table 3.13). The residuals will be further analyzed in the next section. The average probability plot is a graphical representation for determining whether a data set is nearly normally distributed. The normalized residuals are examined on a normal probability map (Figures 3.7 and 3.8) to assess the departure of the data from normality. Observation of the residuals indicates that they are practically in a straight line, revealing the residues that are generally distributed, and the normalcy assumption is true. Figures 3.4 and 3.5 exhibit the histogram plot

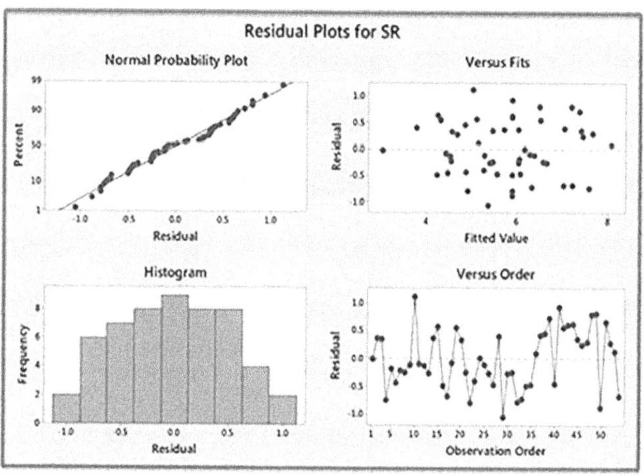

FIGURE 3.4 Residual plot for SR.

TABLE 3.10
Analysis of Variance for SR

Source	DF	Seq SS	Adj SS	Adj MS	F	P
Regression	27	64.426	64.426	2.3861	4.23	0.000
Linear	6	47.588	33.918	5.6530	10.02	0.000
Square	6	5.614	5.614	0.9356	1.66	0.171
Interaction	15	11.224	11.224	0.7483	1.33	0.256
Residual error	26	14.673	14.673	0.5644		
Lack-of-fit	21	11.416	11.416	0.5436	0.83	0.456
Pure error	5	3.258	3.258	0.6515		
Total	53	79.099				

of the standardized residue for all the data; the figure indicates the proportion of the residues. It has a Gaussian distribution (bell shape), and the residues are distributed with a mean of zero. In the calculation, the residual plot versus run order demonstrates that there is no clear pattern or distinctive structure in the data, as illustrated in Figures 3.4 and 3.5. The errors have a constant variance (Table 3.13).

Residual plots are a vital accompaniment to model computations and may be shown against the fitted data to give a visual check on the model assumptions. Experimental data contrasted between the expected values for MRR and SR are presented in Figures 3.2a and 3.2b. The regression model is well-constructed with the given experimental data. It is clearly observed that the experimental and predicted values align closely, forming a straight line, which indicates that the values are suitable for further study.

3.2.5 GRAY RELATIONAL ANALYSIS OPTIMALITY SEARCH

In this study, to renovate the numerous response optimization models into a single response GRA grade, GRA has been applied. Grades have been used to analyse multiresponse characteristics as an alternative by using experimental data directly in GRA and multi-regression models.

3.2.5.1 Steps in GRA

The following steps are taken during GRA in order to determine the grey relational grade and grey relational coefficients given in equations 3.4 and 3.5. By normalizing the MRR and SR experimental results, the effects of accepting various units to reduce variability are avoided.

$$Z_{ij} = \frac{y_{ij} - \min\left(y_{ij}, i = 1,2,\dots n\right)}{\max\left(y_{ij}, i = 1,2,\dots\dots n\right) - \min\left(y_{ij}, i = 1,2,\dots n\right)} \tag{3.4}$$

$$Z_{ij} = \frac{\max\left(y_{ij}, i = 1,2,\dots n\right) - y_{ij}}{\max\left(y_{ij} i = 1,2,\dots\dots n\right) - \min\left(y_{ij}, i = 1,2,\dots n\right)} \tag{3.5}$$

Presenting the grey relational producing and manipulating the grey coefficient for the normalized values yield is given in equation 6.

$$\gamma\left(y_0(k), y_i(k)\right) = \frac{\Delta_{\min} + \xi_{\max}}{\Delta_{oj}(k) + \xi\Delta_{\max}} \tag{3.6}$$

where:
$j = 1, 2 \dots n$; $k = 1, 2 \dots m$, n is the quantity of experimental value and m is the quantity of responses.

y0 (k) is the reference sequence (yo (k) = 1, k = 1, 2 . . . m); yj(k) is the specific comparison sequence.

Δoj = $\|$ yo(k)-yj(k) $\|$ = The absolute value of the difference between y0(k) and yj(k).

Δmin = minmin $\|$ yo(k) – yj(k) $\|$ is the smallest value of yj(k).

Δmax = maxmax $\|$ yo(k) – yj (k) $\|$ is the largest value of yj(k).

ζ is the distinguishing coefficient that is distinct in the range $0 \leq \zeta \leq 1$ (the value may be familiar depending upon the practical needs of the system).

Calculating the GRA by averaging the grey relational coefficient (GRC) is given by:

$$\gamma j = \frac{1}{k} \gamma ij \qquad (3.7)$$

where:

γ_j is the GRA grade assigned to the jth experiment, and k denotes the total number of presentation features. When the original value has the feature of "greater is better", Equation (3.5) is utilised to normalize the experimental result. The following equation is used to normalize MRR. When the novel series exhibits the property "lower is better", the novel progression is normalized using Equation (3.6); that is, SR is normalized using this equation. We calculate the GRC for MRR and SR using Equation (3.7), as given in Table 3.11. Additionally, the GRG is calculated by using Equation (3.8), and the resulting grades are shown in Table 3.11.

ANOVA was created by Sir Ronald Fisher. ANOVA aims to examine the design process parameters in detail, which influences characteristics of quality. Conventional statistical methods focus on one parameter at a time; this analysis is repeated to obtain additional factors for the experiment. Current is a significant element in maximizing the MRR and decreasing the SR, and the ideal design (A1B1C2D2E1F1) is determined from the main effects plot for the LM25 composite in Figure 3.5. After determining the ideal level of EDM machining settings, the next phases validate the development of concert qualities through the optimal combination. Table 3.12 summarizes the results of the experiment, comparing the initial combination machining of EDM parameters to the optimal one. As seen in Table 3.12, the MRR increased from 1.411 to 3.677 mg/sec, while the SR value decreased from 5.09 to 2.91 m. As a result, it is evident that the uniqueness of the exposed quality is expected to be significantly enhanced through this confirmation test.

3.2.6 Performance Enhancement with Particle Swarm Optimization

Particle swarm optimization (PSO) is a well-known advanced technique established by Kennedy and Eberhart. It demonstrates typical evolutionary computation

TABLE 3.11

GRA for LM25 Aluminium Composite

Sl. No.	MRR	SR	Normalized MRR Values	Normalized SR Values	Grey Relational Coefficient of MRR	Grey Relational Coefficient of SR	Grade
1.	1.435	3.01	0.002988792	1	0.994058	0.333333	0.663696
2.	5.8	6.09	0.546575342	0.405405405	0.477749	0.552239	0.514994
3.	4.952	5.82	0.440971357	0.457528958	0.531366	0.522177	0.526772
4.	8.823	6.86	0.923038605	0.256756757	0.351361	0.660714	0.506038
5.	7.88	5.88	0.805603985	0.445945946	0.382965	0.528571	0.455768
6.	2.083	4.43	0.083686177	0.725868726	0.856625	0.407874	0.632249
7.	3.355	4.32	0.242092154	0.747104247	0.673771	0.400929	0.53735
8.	7.96	5.22	0.815566625	0.573359073	0.380064	0.465827	0.422946
9.	6.444	6.27	0.626774595	0.370656371	0.443744	0.574279	0.509012
10.	2.653	6.16	0.154669988	0.391891892	0.763744	0.560606	0.662175
11.	5.205	6.2	0.472478207	0.384169884	0.51415	0.565502	0.539826
12.	1.99	4.41	0.072104608	0.72972973	0.873966	0.406593	0.64028
13.	4.613	5.77	0.39875467	0.467181467	0.556325	0.516966	0.536646
14.	4.511	6.41	0.386052304	0.343629344	0.564301	0.592677	0.578489
15.	2.182	5.69	0.096014944	0.482625483	0.838905	0.508841	0.673873
16.	4.951	5.39	0.440846824	0.540540541	0.531436	0.480519	0.505978
17.	6.059	6.36	0.57882939	0.353281853	0.463465	0.585973	0.524719
18.	2.136	4.35	0.090286426	0.741312741	0.847046	0.402799	0.624923
19.	3.383	4.86	0.245579078	0.642857143	0.67062	0.4375	0.55406
20.	2.206	4.91	0.099003736	0.633204633	0.834719	0.441227	0.637973
21.	5.272	5.79	0.480821918	0.463320463	0.509777	0.519038	0.514407
22.	5.342	4.12	0.489539228	0.785714286	0.505286	0.388889	0.447087
23.	2.28	4.87	0.108219178	0.640926641	0.822072	0.43824	0.630156
24.	4.951	6.17	0.440846824	0.38996139	0.531436	0.561822	0.546629
25.	3.5	5.18	0.26014944	0.581081081	0.657765	0.4625	0.560133
26.	3.248	5.14	0.228767123	0.588803089	0.68609	0.45922	0.572655

(Continued)

TABLE 3.11 *(Continued)*

Sl. No.	MRR	SR	Normalized MRR Values	Normalized SR Values	Grey Relational Coefficient of MRR	Grey Relational Coefficient of SR	Grade
27.	1.411	5.09	0	0.598455598	1	0.455185	0.727592
28.	5.544	7.44	0.514694894	0.144787645	0.492759	0.775449	0.634104
29.	1.906	4.3	0.061643836	0.750965251	0.890244	0.399691	0.644968
30.	7.381	6.39	0.743462017	0.347490347	0.402103	0.589977	0.49604
31.	7.518	6.33	0.760523039	0.359073359	0.396661	0.582022	0.489342
32.	5.272	5.1	0.480821918	0.596525097	0.509777	0.455986	0.482881
33.	6.643	5.6	0.651556663	0.5	0.434195	0.5	0.467097
34.	3.383	3.74	0.245579078	0.859073359	0.67062	0.367898	0.519259
35.	6.343	5.6	0.614196762	0.5	0.448754	0.5	0.474377
36.	6.766	8.19	0.666874222	0	0.428495	1	0.714248
37.	1.684	4.2	0.033997509	0.77027027	0.936334	0.393617	0.664975
38.	2.859	5.31	0.180323786	0.555984556	0.734944	0.473492	0.604218
39.	6.655	8.12	0.653051059	0.013513514	0.433632	0.973684	0.703658
40.	3.866	4.01	0.305728518	0.806949807	0.620556	0.38257	0.501563
41.	4.613	6.82	0.39875467	0.264478764	0.556325	0.65404	0.605183
42.	4.142	7.08	0.340099626	0.214285714	0.595168	0.7	0.647584
43.	4.511	6.49	0.386052304	0.328185328	0.564301	0.60373	0.584015
44.	4.72	6.51	0.412079701	0.324324324	0.548198	0.606557	0.577378
45.	9.441	7.77	1	0.081081081	0.333333	0.860465	0.596899
46.	6.766	7.7	0.666874222	0.094594595	0.428495	0.840909	0.634702
47.	5.486	7.98	0.50747198	0.040540541	0.496292	0.925	0.710646
48.	5.205	8.02	0.472478207	0.032818533	0.51415	0.938406	0.726278
49.	6.444	7.31	0.626774595	0.16988417	0.443744	0.746398	0.595071
50.	3.941	5.01	0.315068493	0.613899614	0.613445	0.448873	0.531159
51.	2.743	4.91	0.165877958	0.633204633	0.750888	0.441227	0.596057
52.	2.985	4.97	0.196014944	0.621621622	0.718375	0.445783	0.582079
53.	2.04	5.28	0.078331258	0.561776062	0.864556	0.470909	0.667733
54.	4.776	6.54	0.419053549	0.318532819	0.544038	0.610849	0.577443

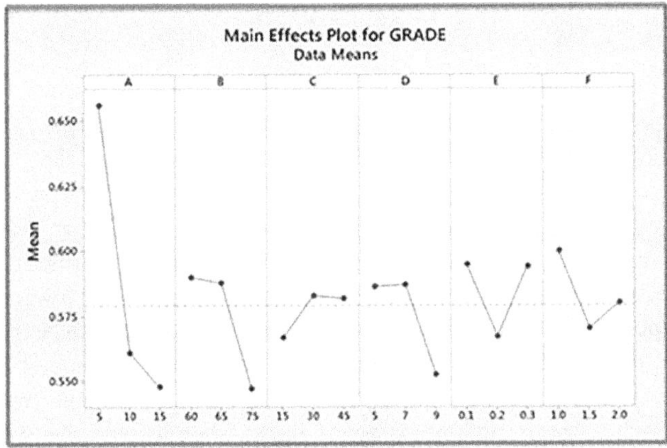

FIGURE 3.5   Main effect plot of input parameters for LM25 Al composite.

TABLE 3.12

Optimal Input and Output Parameters for LM25 Al Composite

	Process Parameters						Output Parameters	
			C	D		F		
	A	B	(s)	(s)	E	Oil		
	(V)	(A)	Pulse-on	Pulse-off	(mm)	Pressure	MRR	
Response	Voltage	Current	Time	Time	Gap	(Kg/cm²)	(Mg/sec)	SR (µm)
Initial	65	5	30	7	0.1	2.0	1.411	5.09
Optimal	60	5	30	7	0.1	1	3.677	2.91

properties, combining population initiation with an arbitrary solution and finding the optimal solution by generation filling. Particles are potential energy solutions. They are "flown" across the crucial gap by the current optimal particles. Each particle that adhered to the designated spatial coordinates achieved an excellent solution. This is referred to as pBest. Another significant value, associated with the global form of PSO, is considered the population's best value, and its position is determined by any particle within the population. This value is referred to as gBest. The PSO conception entails altering the velocity (i.e. acceleration) of each particle towards both its pBest and gBest locations at each step. Figure 3.7 illustrates the step-by-step approach for performing PSO procedures. For both the pBest and gBest locations, random acceleration weights are generated using distinct arbitrary integers. The following equations are used to update the particles:

$$V_{i+1} = w * V_i + c_1 * r_1 * (pBest_i - X_i) + c_2 * r_2 * (gBesti - X_i) \quad (3.8)$$

$$X_{i+1} = X_i + V_{i+1} \qquad (3.9)$$

Equation (3.8) produces an original velocity (Vi+1) for all particles dependent on the previous velocity, together with pBest and gBest. Equation (3.9) offers to update individual particle positions (Xi) in the solution hyperspace.

The random integers r_1 and r_2 in Equation (3.8) are each generated within the range (0, 1).

In Equation (3.9), the acceleration constants c_1 and c_2 represent the weight of the stochastic acceleration functions that pull each particle toward the pBest and gBest places. C_1 reflects the particle's trust in itself (cognitive parameters), whereas C_2 denotes the particle's confidence in the swarm (social parameter). As a result, altering these constants varies the system's average tension value. Constants with a lower value are dragged back, whereas particles with a higher value pass through the target zone. The inertia weight w is has a significant function in the PSO convergence behaviour (Figure 3.6), which is involved in regulating the swarm's inquiry capacities. High inertia weights allow for significant velocity updates, enabling exploration of the design space globally, while low inertia weights restrict velocity updates to nearby regions of the design space. The optimal use of the inertia weight w leads to improved performance across a range of applications. Bergh and Engelbrecht discuss the effects of C1, C2, and w on the convergence of standard numerical objective functions. Table 3.13 shows the optimal value obtained by PSO.

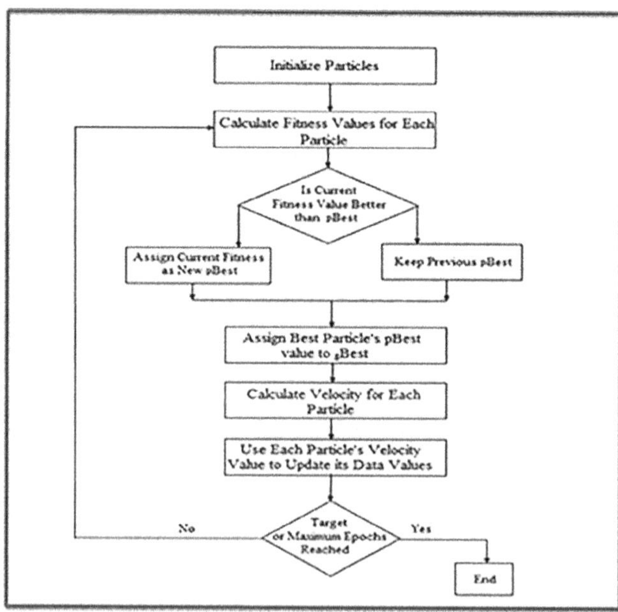

FIGURE 3.6 Flow chart for particle swarm optimization,

TABLE 3.13

Values of MRR and SR for LM25 Al Composite Are Presented through Experimental, Predicted, and Optimized Approaches

Sl. No.	Experimental Value		Prediction RSM		PSO Optimization	
	MRR (Mg/Sec)	SR (µm)	MRR (Mg/Sec)	SR (µm)	MRR (Mg/Sec)	SR (µm)
1.	1.435	3.01	1.344776	3.02	1.427	3.02
2.	5.8	6.09	6.061756	6.16	5.620	6.08
3.	4.952	5.82	5.251514	5.98	4.970	5.42
4.	8.823	6.86	8.973476	6.88	8.920	6.85
5.	7.88	5.88	7.859906	5.8	8.300	5.87
6.	2.083	4.43	1.970906	4.56	2.030	4.32
7.	3.355	4.32	3.160856	4.38	3.530	4.31
8.	7.96	5.22	7.552506	5.33	8.020	5.21
9.	6.444	6.27	6.375656	6.4	6.230	6.25
10.	2.653	6.16	2.740976	6.1	2.710	5.78
11.	5.205	6.2	5.020214	6.03	5.205	6.21
12.	1.99	4.41	2.133056	4.54	1.996	4.32
13.	4.613	5.77	4.523526	5.9	4.730	5.27
14.	4.511	6.41	4.461126	6.3	4.680	5.83
15.	2.182	5.69	2.530314	5.7	2.180	5.68
16.	4.951	5.39	4.661306	5.4	4.954	5.42
17.	6.059	6.36	5.660656	6.53	6.370	6.27
18.	2.136	4.35	2.182026	4.52	2.198	4.56
19.	3.383	4.86	3.450356	4.88	3.470	4.87
20.	2.206	4.91	2.489656	4.9	2.200	4.75
21.	5.272	5.79	5.139914	5.8	5.270	5.98
22.	5.342	4.12	5.155064	4.2	5.570	4.13
23.	2.28	4.87	1.991506	4.9	2.380	4.86
24.	4.951	6.17	4.891926	6.01	4.950	6.02
25.	3.5	5.18	3.335956	5.23	3.350	4.91
26.	3.248	5.14	3.526756	5.22	3.520	5.1
27.	1.411	5.09	1.014914	5.09	1.507	5.1
28.	5.544	7.44	5.948156	7.2	5.600	7.32
29.	1.906	4.3	1.677426	4.19	2.010	4.02
30.	7.381	6.39	7.410814	6.44	6.730	6.17
31.	7.518	6.33	7.071526	6.52	7.430	6.3
32.	5.272	5.1	4.661306	5.2	5.340	4.85
33.	6.643	5.6	6.345656	5.68	6.780	5.61
34.	3.383	3.74	2.860256	3.79	6.674	2.78
35.	6.343	5.6	6.120556	5.8	6.370	5.61
36.	6.766	8.19	6.260856	8.23	6.920	7.92
37.	1.684	4.2	1.617564	4.3	1.687	4.13

(Continued)

TABLE 3.13 *(Continued)*

Sl. No.	Experimental Value		Prediction RSM		PSO Optimization	
	MRR (Mg/Sec)	SR (µm)	MRR (Mg/Sec)	SR (µm)	MRR (Mg/Sec)	SR (µm)
38.	2.859	5.31	3.058856	5.34	2.850	5.2
39.	6.655	8.12	6.959414	8.2	6.650	8.02
40.	3.866	4.01	3.827276	4.04	3.870	4.02
41.	4.613	6.82	4.661306	6.9	4.620	6.83
42.	4.142	7.08	4.493526	7.23	4.320	7.07
43.	4.511	6.49	4.661306	6.33	4.511	6.45
44.	4.72	6.51	4.661306	6.53	4.720	6.55
45.	9.441	7.77	9.501460	7.624	9.533	7.63
46.	6.766	7.7	6.762056	7.9	6.720	7.71
47.	5.486	7.98	5.907056	7.99	5.480	7.57
48.	5.205	8.02	5.502926	8.01	5.205	8.03
49.	6.444	7.31	6.790656	7.21	6.570	7.03
50.	3.941	5.01	4.661306	5.23	4.020	4.68
51.	2.743	4.91	3.172056	4.8	2.870	4.72
52.	2.985	4.97	3.204856	4.8	3.040	4.86
53.	2.04	5.28	2.153964	5.23	2.143	5.32
54.	4.776	6.54	4.572614	6.55	4.776	5.93

3.3   TEST RESULTS AND DISCUSSION

3.3.1   INFLUENCE OF MACHINING PARAMETERS ON MRR

Experiments are conducted to determine the efficiency of machining parameters (V_d, T_{off}, T_{on}, I_p, P_{oil}, and S_g) on the response variables MRR. The solution findings were further investigated using the Minitab software by sharing the data. The second-order model's objective was to ascertain the link between process variables and the MRR. ANOVA was performed to determine the second-order model's adequacy. Figure 3.7 illustrates the computed response surface of the MRR in relation to the pulse parameters T_{on} and I_p.

The chart clearly demonstrates that the MRR causes a significant increase in I_p for a given amount of T_{on} time. Thus, increased MRR is obtained at high I_p values and long T_{on} time due to their leading restriction to the input energy; that is, the increase in pulse current generates a strong spark, which generates a high temperature, which causes many materials in the work piece to erode and melt.

The effect of T_{off} and I_p on the observed MRR response surface is seen in Figure 3.7b; the parameters V_d, T_{on}, S_g, and P_{oil} remain constant at their maximum values. It is observed that the MRR increases as the I_p increases. As previously stated, when T_{off} increases, MRR decreases; this is due to T_{off} increases, indicating an unspecific temperature loss that does not contribute to MRR. This has

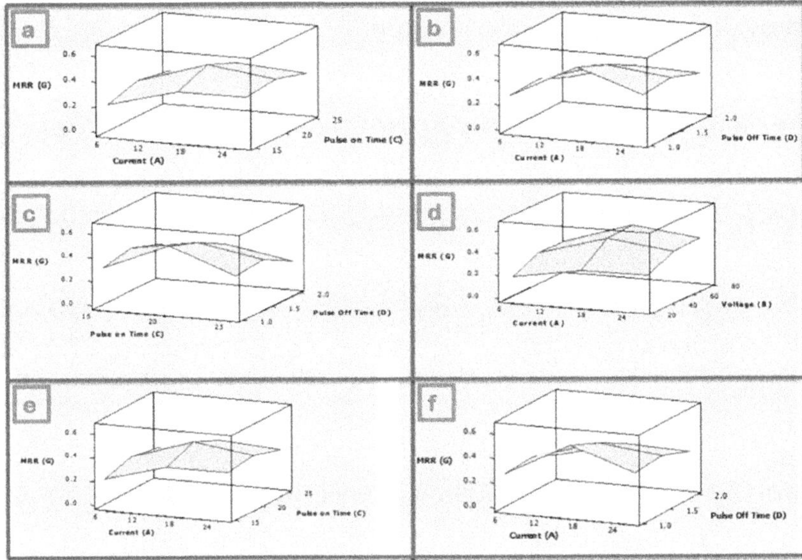

FIGURE 3.7   MRR vs a) current and pulse-on time; b) current and pulse-off time; c) pulse-on time and pulse-off time; d) current and voltage; e) current and pulse-on time; f) current and pulse-off time.

the effect of bringing the temperature of the work piece to a halt prior to the second spark initiating, lowering the MRR. For a given set of input parameters, the enhanced MRR is attained with a high I_p and a low T_{off}. MRR is shown as a function of T_{on} and T_{off} in Figure 3.7c, but I_p, V_d, S_g, and P_{oil} remain steady at their higher levels. Through monitoring, the significant MRR values are assigned to the greater T_{on} range and the lower T_{off} range. MRR as a function of V_d and I_p is shown in Figure 3.7d, whereas T_{on}, T_{off}, S_g, and P_{oil} remain constant at the higher level. As a result of the observation, it is clear that high MRR values occur at high I_p and V_d. MRR is shown as a function of I_p and T_{on} in Figure 3.7e, while I_p, voltage, T_{off}, S_g, and P_{oil} remain constant at their high-range values. According to observation, the highest MRR occurred as a result of increased T_{on} and I_p. MRR is shown as a function of I_p and T_{off} in Figure 3.7f, whereas I_p, T_{on}, V_d, S_g, and P_{oil} remain constant in the higher order. It is noted that the high MRR happened as a result of the high I_p range and low T_{off}.

Figure 3.8a depicts MRR as a function of V_d and P_{oil}, whereas T_{on}, S_g, T_{off}, and I_p remain stable in their higher degrees of order. It is found that increased MRR values occurred with the higher V_d and medium P_{oil}. Figure 3.8b shows MRR as a function of T_{off} and P_{oil}, while I_p, S_g, T_{on}, and V_d remain steady at a high level. It is observed that the highest MRR values occur at lower values of T_{off} and P_{oil}. Figure 3.8c displays MRR as a function of S_g and V_d, while I_p, T_{off}, T_{on}, and I_p remain stable at their increased levels. The insight is that the largest MRR values occurred at smaller S_g and higher V_d. Figure 3.8e displays MRR as a function

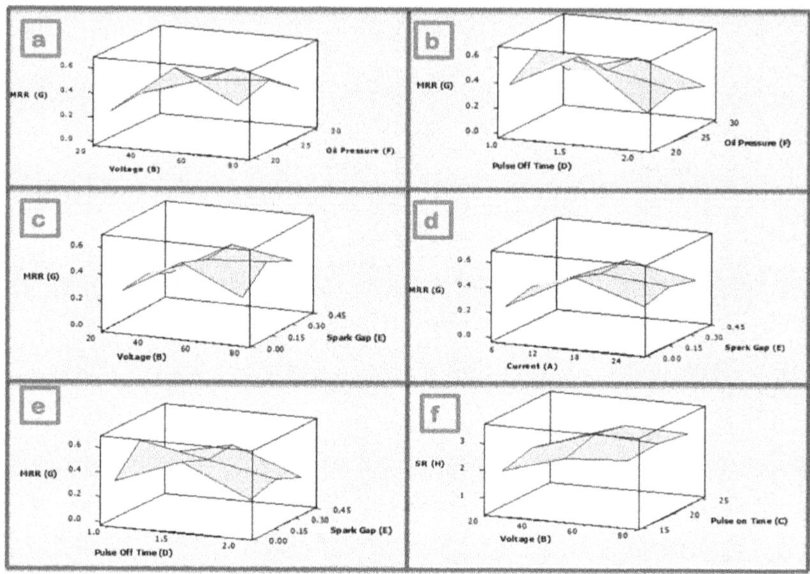

FIGURE 3.8 MRR vs a) voltage and oil pressure; b) pulse-on time and oil pressure; c) voltage and spark gap; d) current and spark gap; e) pulse-off time and spark gap; f) SR vs voltage and pulse-on time.

of S_g and I_p, whereas the voltage, T_{off}, T_{on}, and P_{oil} stay stable at a higher level. It can be proved that the largest MRR values happened at lower S_g and higher I_p. Figure 3.8f depicts MRR as a function of S_g and T_{off}, but I_p, T_{on}, P_{oil}, and V_d remain constant at a higher degree of order. It is observed that the biggest MRR values occurred at smaller S_g and lower T_{off} time.

3.3.2 INFLUENCE OF MACHINING PARAMETERS ON SR

The effectiveness of the machining process parameters (I_p, V_d, T_{off}, T_{on}, S_g, and P_{oil}) on the response variable SR is evaluated through conducted experiments. The results were sent to Minitab for additional research. The second-order model was utilized to investigate the association among MRR and the process variables taken into account. ANOVA was employed to assess the appropriateness of the second-order model. Figure 3.9a displays SR as a function of V_d and T_{on}, whereas T_{on}, I_p, S_g, and P_{oil} stay constant at their lower values. It has been noted that the SR values are significantly diminished when V_d and T_{on} are both lowered. Figure 3.9b shows SR as a function of T_{off} and V_d, although I_p, T_{on}, S_g, and P_{oil} stay stable at lower values. It is noted that the SR values are significantly lower when both the V_d and T_{on} are low. Figure 3.9c depicts SR as a function of T_{off} and T_{on}, although V_d, I_p, S_g, and P_{oil} stay constant at their lower values. It is observed that SR values are lower when the T_{on} is reduced and the T_{off} is increased. However, these two factors have a minimal effect compared to the influence of Ip on SR.

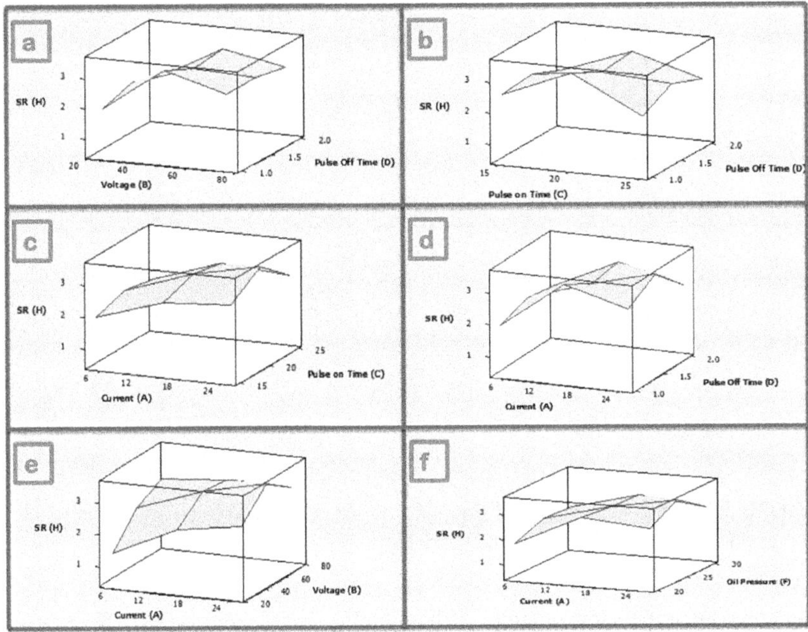

FIGURE 3.9   SR vs a) voltage and pulse-off time; b) pulse-on time and pulse-off time; c) current and pulse-on time; d) current and pulse-off time; e) current and voltage; f) voltage and oil pressure.

Figure 3.10d illustrates the measured response surface for SR in relation to the process parameters I_p and T_{on}, with T_{off}, V_d, S_g, and P_{oil} remaining constant at their minimum values. As seen in the figures, raising I_p results in a significant increase in the SR for any value of T_{on}. Conversely, the SR results in higher values with maximum T_{on}, especially when the I_p is increased. Thus, the smallest SR is obtained with a lower I_p and a shorter T_{on} time. Due to their leading constraint on the input energy, increasing I_p generates a powerful spark, creating a higher temperature and crater, resulting in a rough surface on the work piece, while decreasing I_p generates a tiny crater and hence a smooth surface. Figure 3.10e illustrates the effect of T_{off} and I_p on the measured response of SR. At lower levels, T_{on}, V_d, S_g, and P_{oil} stay constant. By demonstrating that an increase in SR results in an increase in I_p and T_{off}. SR is shown as a function of I_p and V_d in Figure 3.10f, whereas T_{on}, S_g, T_{off}, and P_{oil} stay constant at the lower-level order. This demonstrates that SR values are low when both the V_d and I_p are small. SR is shown in Figure 3.10 a as a function of P_{oil} and I_p, with T_{on}, V_d, T_{off}, and S_g remaining constant at lower levels. It is observed that SR values are lower when both the P_{oil} and I_p values are low. Figure 3.10b illustrates the influence of P_{oil} and T_{off} on the predictable response surface of SR. T_{on}, I_p, V_d, and S_g all stay constant at their lowest values. The SR value rises as P_{oil} increases and the T_{on} time decreases. SR is shown as a function of P_{oil} and V_d in Figure 3.10c, although the T_{on}, I_p, T_{off}, and S_g

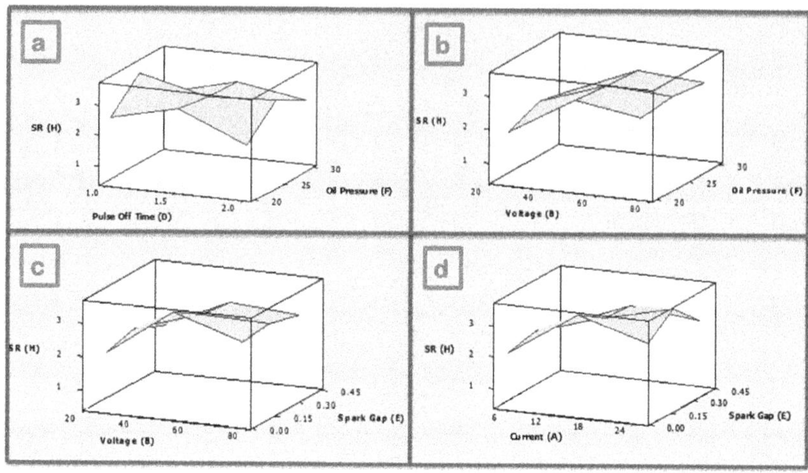

FIGURE 3.10 SR vs a) pulse-off time and oil pressure; b) voltage and oil pressure; c) voltage and spark gap; d) current and spark gap.

stay constant at their lower values. When the P_{oil} and V_d are low, the SR values are lower. SR is shown as a function of V_d and S_g in Figure 3.10d, but I_p, T_{on}, T_{off}, and P_{oil} stay constant at their lower values. It is noticed that when V_d and S_g are high, the SR values are lower. SR is shown as a function of S_g and I_p in Figure 3.10e, while T_{on}, V_d, T_{off}, and P_{oil} stay constant at their lower values.

When observing smaller values for S_g and I_p, a noticeable trend emerges wherein the SR values tend to decrease.

Comparison and validation of RSM and PSO results with experimental data proved the high accuracy of models, as shown in Figure 3.11a for MRR and Figure 3.11b for SR. PSO optimization methodology might be an affordable and effective solution for optimization of EDM and output parameters according to input variables.

As a result, this technique should significantly enhance the EDM process outcomes in terms of SR and MRR. The greatest influencing element for MRR and SR is the current. The ideal input and output for this test are shown in Table 3.14.

The individual value plots using the t-test for the PSO and experimental results are shown in Figure 3.12. The paired t- test is used to test the significance of the results. Since the p-values are less than 0.05, the PSO results are statistically significant.

3.4 COMPARISON OF RESULTS

RSM-PSO was combined to provide a good response in acquiring optimum parametric solutions. The 60 V discharge voltage, 10 A current, 15 S T_{off}, 7 S T_{on}, 0.2-mm spark gap, and 1 kg/cm² oil pressure give a maximum MRR with minimum SR. The optimum MRR obtained from the PSO method is 3.37 times higher

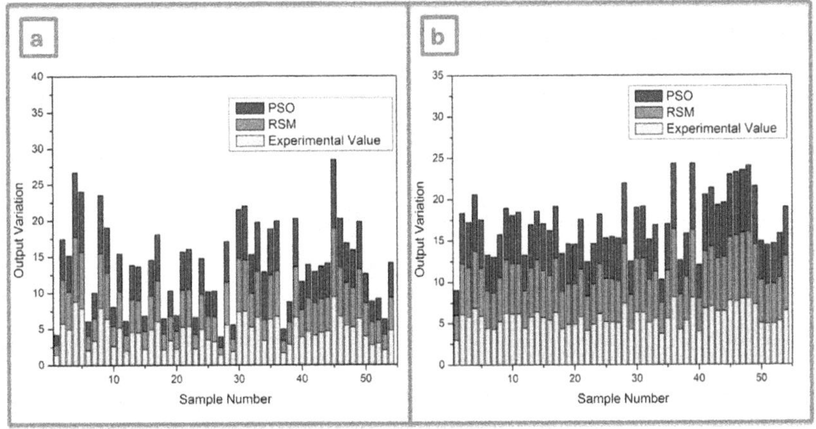

FIGURE 3.11 Comparison chart for a) MRR; b) SR.

TABLE 3.14
Optimal Input and Output Parameters for LM25 Al Composite

		Process Parameters				Output Parameters	
A **(V)** **Voltage**	**B** **(A)** **Current**	**C** **(s)** **Pulse-on** **Time**	**D** **(s)** **Pulse-off** **Time**	**E** **(mm)** **Gap**	**F** **Oil** **Pressure** **(Kg/cm²)**	**MRR (Mg/** **sec)**	**SR (μm)**
60	10	15	7	0.2	1.0	6.674	2.78

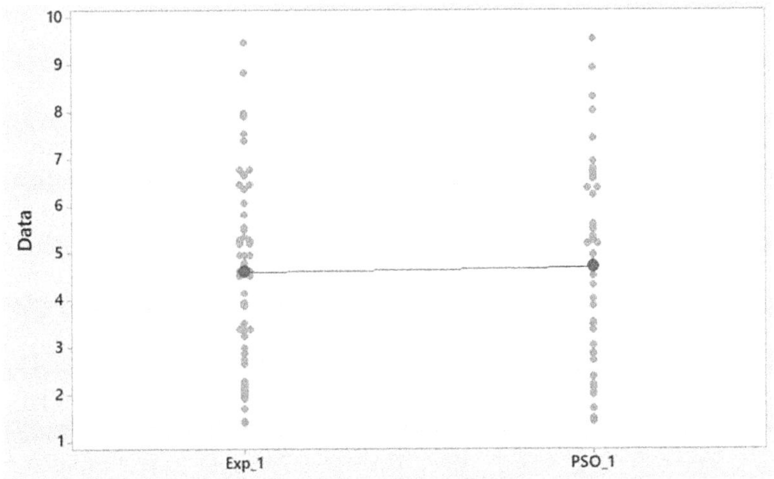

FIGURE 3.12 Comparison of t-test results for MRR.

TABLE 3.15

Effectiveness of the Optimization Methods

Datasets	RMSE	MSE	MAPE	MAE
RSM	3.0827	9.5028	19.45	1.5511
Grey method	3.6337	13.204	16.723	1.5158
PSO	2.2282	4.965	13.066	1.1232

TABLE 3.16

Analysis in Terms of Computational Time

Methods	Time (sec)
RSM	1910.8
Grey method	1709.7
PSO	1504.3

compared to the initial value and 2.99 times higher than RSM values. Also, SR is reduced to 45.38% from the initial value and has a 4.67% reduction as compared to RSM method.

The effectiveness of the PSO method is compared with RSM and the grey method and is shown in Table 3.15. The analysis in terms of computational time is given in Table 3.16.

3.5 CONCLUSION

This study delves into the modelling and optimization aspects of the EDM process parameters. Employing the RSM, mathematical models were formulated to effectively manage the combined influence of peak current, discharge voltage, pulse-off-time, pulse-on-time, and oil pressure on both MRR and SR. Optimal process parameter settings were attained using a combination of RSM optimization charts, mathematical modelling, GRA, and PSO techniques. The validity of the second-order response models was verified through an in-depth analysis of variance (ANOVA) for both MRR and SR. Furthermore, the study aimed to forecast the ideal machining conditions for achieving the highest MRR and lowest SR. This was accomplished by leveraging optimization approaches, mathematical modelling, and PSO. The identified optimal conditions for achieving maximum MRR and minimum SR were found to be T_{on} (15 s), T_{off} (7 s), V_d (60 V), I_p (10 A), S_g (0.2 mm), and P_{oil} (1.0 kg/cm²) for pulse-on-time, pulse-off-time, input voltage, current, spark gap, and oil pressure, respectively.

REFERENCES

1. Chaubey S K and Jain N K 2019 Analysis and multi-response optimization of gear quality and surface finish of meso-sized helical and bevel gears manufactured by WSEM process. *Precision Engineering*. 55 (5): 293–309.

2. Gupta K and Jain N K 2014 Analysis and optimization of micro-geometry of miniature spur gears manufactured by wire electric discharge machining. *Precision Engineering*. 38 (4): 728–37.

3. Kanlayasiri K and Jattakul P 2013 Simultaneous optimization of dimensional accuracy and surface roughness for finishing cut of wire-EDMed K460 tool steel. *Precision Engineering*. 37 (3): 556–61.

4. Wang P J and Tsai K M 2001 Semi-empirical model on work removal and tool wear in electrical discharge machining. *Precision Engineering*. 114 (1): 1–17.

5. Valentincic J and Junkar M 2004 A model for detection of the eroding surface based on discharge parameters. *Machine Tools and Manufacture*. 44 (2–3): 175–81.

6. Jaharah A G, Liang C G, Wahid S Z, AbRahman M N and Che Hassan C H 2008 Performance of copper electrode in electrical discharge machining (EDM) of AISI H13 harden steel. *Mechanical and Materials Engineering*. 3 (1): 25–29.

7. Kanagarajan D, Karthikeyan R, Palanikumar K and Sivaraj P 2008 Influence of process parameters on electric discharge machining of WC/30%Co composites. *Engineering Manufacture*. 222 (7): 807–15.

8. Kuppan P, Rajadurai A and Narayanan S 2008 Influence of EDM process parameters in deep hole drilling of Inconel. *Advanced Manufacturing Technology*. 38 (1–2): 74–84.

9. Puertas I and Luis C J 2004 A study of optimization of machining parameters for electrical discharge machining of boron carbide. *Materials and Manufacturing Processes*.19 (6): 1041–70.

10. Khan A A 2008 Electrode wear and material removal rate during EDM of aluminum and mild steel using copper and brass electrodes. *Journal of Advanced Manufacturing Technology*. 39 (5–6): 482–87.

11. Khan A A, Ali M Y and Haque M M 2009 A study of electrode shape configuration on the performance of die sinking EDM. *Mechanical and Materials Engineering*. 4 (1): 19–23.

12. Dhar S, Purohit R, Saini N, Sharma A and Kumar G H 2007 Mathematical modeling of electric discharge machining of cast Al-4Cu-6Si alloy-10 wt.% SiCP composites. *Materials Processing Technology*. 194 (1–3): 24–29.

13. Salonitis K, Stournaras A, Stavropoulos P and Chryssolouris G 2009 Thermal modeling of the material removal rate and surface roughness for die-sinking EDM. *Advanced Manufacturing Technology*. 40 (3–4): 316–23.

14. El-Taweel T A 2009 Multi-response optimization of EDM with Al-Cu-Si-TiC P/M composite electrode. *Advanced Manufacturing Technology*. 44 (1–2): 100–13.

15. Chiang K T 2008 Modeling and analysis of the effects of machining parameters on the performance characteristics in the EDM process of Al_2O_3+TiC mixed ceramic. *Advanced Manufacturing Technology*. 37 (5–6): 523–33.

16. Akhay D, Kumar P and Inderdeep S 2008 Experimental investigation and optimisation in EDM of Al 6063 SiCp metal matrix composite. *Machining and Machinability of Materials*. 3 (5): 293–308.

17. Karthikeyan R, Lakshmi Narayanan P R and Naagarazan R S 1999 Mathematical modelling for electric discharge machining of aluminium-silicon carbide particulate composites. *Materials Processing Technology*. 87 (1–3): 59–63.

18. Wang X, Han P, Giovannini M and Ehmann K 2017 Modeling of machined depth in laser surface texturing of medical needles. *Precision Engineering.* 47: 10–18.
19. Mohri N, Takezawa H, Furutani K, Ito Y and Sata T 2000 New process of additive and removal machining by EDM with a thin electrode. *CIRP Annals—Manufacturing Technology.* 49 (1): 123–26.
20. Marafona J and Wykes C 2000 New method of optimising material removal rate using EDM with copper-tungsten electrodes. *Machine Tools and Manufacture.* 40 (2): 153–64.
21. Bleys P, Kruth J P, Lauwers B, Zryd A, Delpretti R and Tricarico C 2002 Real-time tool wear compensation in milling EDM. *CIRP Annals—Manufacturing Technology.* 51 (1): 157–60.
22. Kunieda M and Kobayashi T 2004 Clarifying mechanism of determining tool electrode wear ratio in EDM using spectroscopic measurement of vapor density. *Materials Processing Technology.* 149 (1–3): 284–88.
23. Habib S S 2009 Study of the parameters in electrical discharge machining through response surface methodology approach. *Applied Mathematical Modelling.* 33 (12): 4397–407.
24. Staelens F 1989 A computer integrated machining strategy for planetary EDM. *CIRP Annals—Manufacturing Technology.* 38 (2): 187–90.
25. Yu Z Y, Masuzawa T and Fujino M 1998 Micro-EDM for three-dimensional cavities—development of uniform wear method. *CIRP Annals—Manufacturing Technology.* 47 (1): 169–72.
26. MahdaviNejad R A 2011 Modelling and optimization of electrical discharge machining of SiC parameters, using neural network and non-dominating sorting genetic algorithm (NSGA II). *Materials Sciences and Applications.* 2 (6): 669–75.
27. Yilmaz O, Eyercioglu O and Gindy N N Z 2006 A user-friendly fuzzy-based system for the selection of electro discharge machining process parameters. *Materials Processing Technology.* 172 (3): 363–71.
28. Pritam C, Bhausaheb D Y, Gautam R, KiranNaik B and Kumar S V 2023 Parametric optimization of wire EDM process for single crystal pure tungsten using Taguchi-Grey relational analysis. *Sadhana* 48 (152): 1–10.
29. Çakıroğlu R 2022 Analysis of EDM machining parameters for keyway on Ti-6Al-4V alloy and modelling by artificial neural network and regression analysis methods. *Sadhana.* 47 (150): 1–17.

4 Analysis on Hard Turning of Oil-Hardened Non-Shrinking Die Steel Using the Taguchi Method with Decision Tree Algorithms

Jafrey Daniel James D, Balasundaram R, Karthik Pandiyan G, and Muthu Chozha Rajan B

4.1 HARD TURNING

Hard turning is a process that is defined as turning material with a hardness more than 45 Rockwell hardness C scale. The hard turning of material is carried out with cutting tools made up of cubic boron nitride, ceramics or specially coated tools (Singh and Rao 2007). In certain cases, a large number of combined operations are required for finishing a product. Hard turning is becoming a common practice in several industries nowadays due to an increase in production rates (Koenig et al. 1984).

Some of the advantages of hard turning are an increase in productivity, reduction in setup times and better surface roughness when compared with grinding. The machine used for hard turning must be rigid, with adequate power, hard tool materials with proper geometry and tool holders. The tools developed initially for the cutting of hard materials are alumina-based tools and silicon nitride–based tools. In recent years, special coated tools or grooved of tools have been used for hard turning of materials (Tonshoff et al. 2000).

The selection of process parameters is an essential factor during the hard turning process. It can be selected based on handbooks, which are not always correct. Factors like surface roughness, force during cutting, consumption of power and wear rate of the tool are the deciding factor of surface roughness. Surface roughness during hard turning is influenced by a number of factors like feed rate during

DOI: 10.1201/9781003397465-4

"

machining, speed of cutting, hardness of work piece, time consumed for cutting and cutting tool geometry (Vereschaka et al. 2019).

4.2   DECISION TREE ALGORITHMS

In recent years, all production processes have used a combination of experimental and computational methodologies to reduce the number of experiments while enhancing the production process (Balachandhar et al. 2021). Machine learning methods can be used to investigate the data sets' complicated interdependencies (Huber et al. 2020). One of the most actively investigated fields in machine learning is decision trees (DTs). Aside from the capacity to explain the decision-making process and minimal processing costs, decision trees typically produce good outcomes when compared to other machine learning techniques. Despite the fact that the most widely used decision tree induction techniques, like C4.5 and CART, were invented some time ago, they are still widely utilized to solve common classification tasks (Katz et al. 2014).

A decision tree is a form of directed graph that is used to sort things into categories. It is made up of a root node (a graph node to which no other node points), internal nodes (nodes that point to other nodes and are directed at) and leaves (nodes that do not point to other nodes). The categorized object "travels" from the root to one of the leaves throughout the classification process, where a classification is made. One of the available classes or a collection of probabilities can be used to classify anything (one for each of the class values that could be used) (Ahmed et al. 2018).

There are several algorithms for producing decision trees, as well as a variety of techniques for doing so. All decision tree methods involve recursive partitioning, but they differ in how they select the criterion by which to split each node in the tree and the halting criteria (the decision not to split a node any further). The ID3 and C4.5 algorithms use information gain (IG) and gain ratio measures, but the tests are performed on basis of attributes. In order to provide the method's prediction, the CART algorithm uses the Gini impurity measure and regression in the leaves (Govindan et al. 2017). A decision tree algorithm (C 4.5 algorithm) is used for the L9 orthogonal array in this work, which is motivated by the strength of the data mining for the production process.

Based on a broad literature review, it was discovered that there are few works on the hard turning of OHNS materials. The work began with heat treatment of OHNS material. The experiments were designed using a L9 orthogonal array with feed rate, speed and DoC as input process parameters. The output measured was material removal rate, surface roughness and machining time. Finally, decision tree algorithms were used to study the effect of the process parameters.

4.3   MATERIALS SELECTED AND METHOD

The following sections go over the material selection, machining process parameters and output response calculation.

4.4 OHNS

OHNS was the material used in this study, and its chemical composition is listed in Table 4.1. It was hardened at temperatures ranging from 790°C to 820°C. The OHNS bars in a cylindrical shape were machined in dry conditions. The OHNS material was trued and cleaned prior to machining experiments by extracting 0.3 mm DoC from the outer surface (James Dhilip et al. 2020).

4.5 EXPERIMENTATION

The OHNS material was machined using a commercially available carbide insert (Carbide-TAEGUTEC-TT-5100-04). The carbide insert was coated by chemical vapor deposition method. It was coated with TiCN-Al2O$_3$-TiN (James Dhilip et al. 2020). The Taguchi method was used to design the experiments. The experiment's control parameters are the feed rate, speed, and DoC. Each level was classified into three parameters, as shown in Table 4.2. A L9 orthogonal array was used to conduct the experiment. Surface roughness, machining time and MRR were the output characteristics studied. Equation 4.1 was used to calculate the MRR. Surface roughness was calculated using a surface roughness tester.

$$MRR = \frac{\textit{Weight of OHNS material before machining} - \textit{Weight of OHNS material after machining}}{\textit{Machining time}} \qquad (4.1)$$

TABLE 4.1

Chemical Composition of OHNS Material

Typical Chemical Composition

C	Si	Cr	Mn	S
0.85–0.95%	0.15–0.40%	0.2–0.5	1.8–2.2%	0.05%

TABLE 4.2

Process Parameters and Their Levels

	Process Parameters and Levels		
Levels	Speed (rpm)	Feed (mm/rev)	DoC
1	1500	0.10	0.1
2	1750	0.15	0.2
3	2000	0.20	0.3

4.6 EXPERIMENTAL OUTPUT

Table 4.3 displays the results of the experiments carried out, and the main effect plots are shown in Figure 4.1(a–c). When the speed is greater than or equal to 2000 rpm, the machining time is at its maximum. Generally, the machining speed was low when there was an increase in speed. When considering feed, the optimum was when the feed was less than or equal to 0.2 mm/rev. When the DoC was considered, it decreased, and it increased on further increase in the DoC by 0.3. The main

TABLE 4.3

Input and Output Values for Turning of OHNS

Sl. No	Speed (rpm)	Feed (mm/rev)	DoC	MT (min)	Ra (µm)	MRR
1	1500	0.1	0.1	1.52	0.67	1.5
2	1500	0.15	0.2	1.41	1.064	0.742
3	1500	0.2	0.3	1.38	0.98	1.348
4	1750	0.1	0.2	1.42	0.63	1.246
5	1750	0.15	0.3	1.36	0.555	1.282
6	1750	0.2	0.1	1.29	1.231	1.082
7	2000	0.1	0.3	1.32	0.65	1.85
8	2000	0.15	0.1	1.28	0.865	1.453
9	2000	0.2	0.2	1.17	1.602	1.193

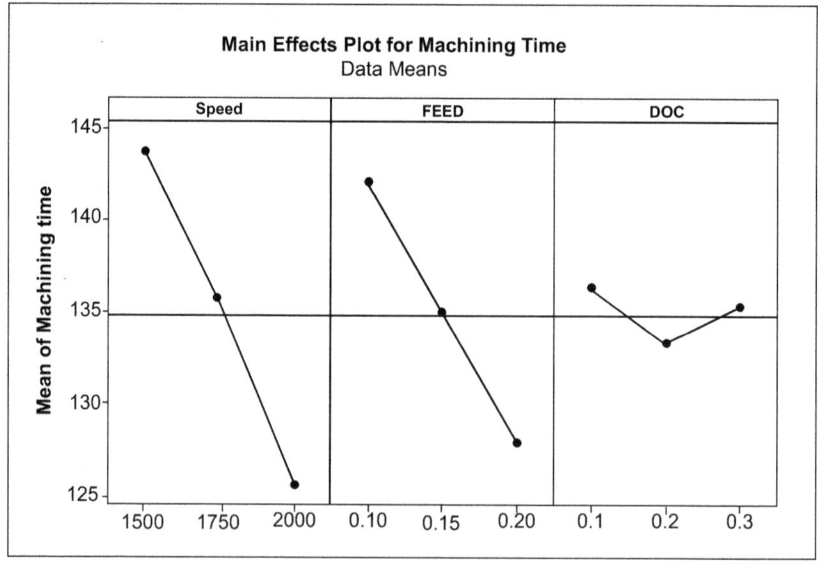

FIGURE 4.1(A) Main effect plots of machining time.

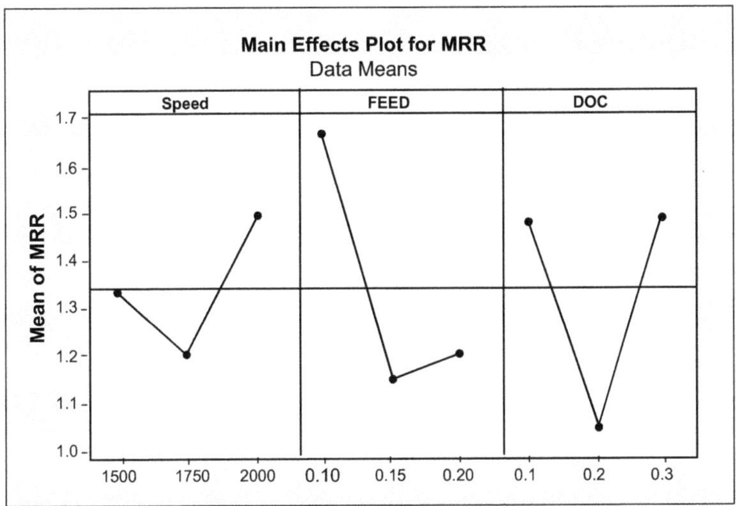

FIGURE 4.1(B) Main effect plots of MRR.

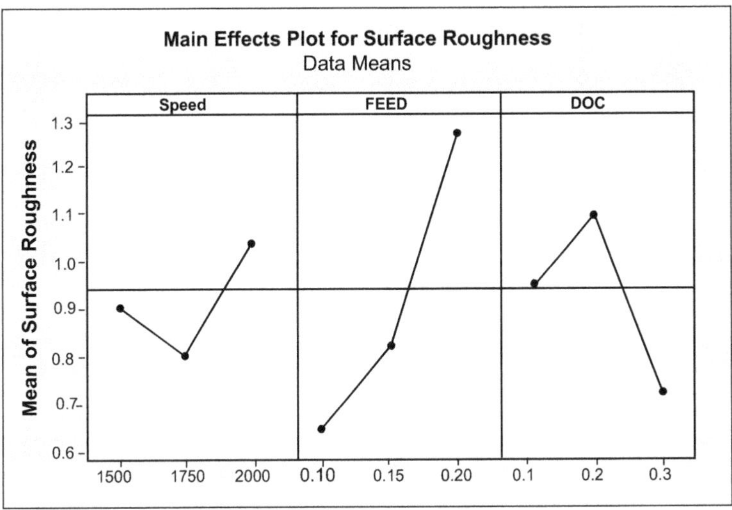

FIGURE 4.1(C) Main effect plots of surface roughness.

effect plots revealed that when the speed exceeded 1500 rpm, there was an increase in MRR due to plastic deformation. The MRR was low when the feed rate was increased. When the DoC increased, the MRR increased as well. According to the previous conclusion, in the case of MRR, cutting speed was the most important factor. The surface roughness was good when the speed was lower, but as the speed increased, there was a decrease in surface roughness. Surface roughness was superior when the feed rate was less than or equal to 0.15 mm/rev and low when the feed rate was greater than 0.15 mm/rev. Surface roughness was better when the DoC was less than or equal to 0.2 mm (James Dhilip et al. 2020).

4.7 DECISION TREE ALGORITHM FOR OUTPUT PARAMETERS

The supervised machine learning method decision tree is concerned with the link between a collection of known input and output qualities. It is one of the most effective methods since it is simple to use and the principles are easy to grasp and comprehend. A decision tree algorithm could be used to create a decision tree T from the collection of input attributes. Quinlan developed the Dichotomiser 3 (ID3) algorithm for building decision trees, and C4.5 is the improved version. C 4.5 is used in the current investigation, using predictor variables as input and class variables as output. The average of each output value in Table 4.3 was used as a baseline for transforming variables into categorical ones. A value less than the average was considered "low," while a value greater than the average was considered "high." Table 4.3 has been reconstructed according to this, and the values are shown in Table 4.4.

4.8 IMPLEMENTATION OF DECISION TREE ALGORITHM FOR MACHINING TIME

Entropy (IG) has been estimated in order to construct the decision tree for machining time. Claude E. Shannon established entropy as the value of IG, as shown in Equation 4.2.

$$Entropy\left(P_1, P_2 \ldots P_N\right) = -P_1 log_2 P_1 - P_2 log_2 P_2 . - \ldots\ldots - P_n log_2 P_n \quad (4.2)$$

TABLE 4.4

Experimental Values of L9 Orthogonal Array for Construction of DT

	Input Variables (Predictor Attributes)			Output Variables (Class Attributes)		
Instance No.	Speed (rpm)	Feed (mm/rev)	DoC (mm)	Machining Time (min)	Surface Finish	MRR
1	1500	0.1	0.1	High	Good	High
2	1500	0.15	0.2	High	Good	Low
3	1500	0.2	0.3	High	Good	High
4	1750	0.1	0.2	High	Good	Low
5	1750	0.15	0.3	High	Good	Low
6	1750	0.2	0.1	Low	Bad	Low
7	2000	0.1	0.3	Low	Good	High
8	2000	0.15	0.1	Low	Good	High
9	2000	0.2	0.2	Low	Bad	Low

The steps which are followed for implementing a decision tree algorithm to frame the rules for calculating machining time are as follows.

4.8.1 CONSTRUCTION OF ROOT NODE

The input attribute (i.e. input machining conditions) is considered the root node.

4.8.2 CALCULATION OF THE CLASS ATTRIBUTE'S INFORMATION GAIN

Based on some conditions, the attributes for machining time in Table 4.2 are grouped as high and low values. In this case, the number of high (H) values is 5, and the number of low values (L) is 4.

From Equation 4.2:

$$Info(5,4) = -\frac{5}{9} log_2 \frac{5}{9} - \frac{4}{9} log_2 \frac{4}{9}$$

$$= (0.471) + (0.519) = \mathbf{0.99}$$

The total gain or entropy function of the L9 orthogonal array for machining time is shown in Info (5, 4).

4.8.4 CALCULATION OF INFORMATION GAIN AND TREE STRUCTURE OF ATTRIBUTES OF PREDICTOR (IG FOR THE ATTRIBUTE SPEED

For calculating the IG of the attribute speed, Table 4.5 and Figure 4.2 illustrate the number of low and high class values for corresponding instance numbers.

TABLE 4.5
Number of Class Attributes for Various Speed Values

Speed	No. of Low (L)	No. of High (H)
1500	0	3
1750	1	2
2000	3	0

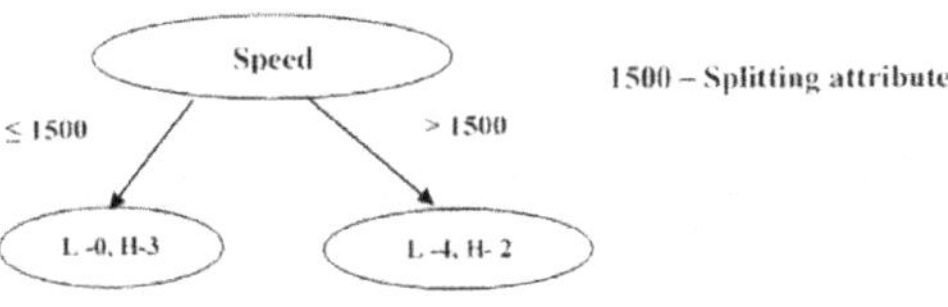

FIGURE 4.2 Decision tree for attribute speed.

The tree can be constructed using either 1500 rpm or 2000 rpm. Since it has both cases, one of the attributes has a value of zero. Here speed is taken as 1500 rpm, and the tree is constructed.

From Equation 4.2:

$$Info(0,3) = -\frac{3}{3}\log_2\frac{3}{3} = \mathbf{0}$$

$$Info(4,2) = -\frac{4}{6}\log_2\frac{4}{6} - \frac{2}{6}\log_2\frac{2}{6} = \mathbf{0.917}$$

$$Combined\ Info\big((0,3),(4,2)\big) = \frac{3}{9}\times(0) + \frac{6}{9}\times 0.917 = \mathbf{0.611}$$

$$Gain\ of\ attribute = \big(overall\ gain\ class\ attribute\big)$$
$$-\big(combined\ info\ of\ reinforcement\big) \qquad (4.3)$$

From Equation 4.3:

$$Gain\ of\ attribute\ speed = (0.99) - (0.611) = \mathbf{0.378}$$

1. **IG for the attribute feed**

   To calculate the IG of the feed attribute, Table 4.6 and Figure 4.3 illustrate the number of low and high class values for corresponding instance numbers.

   From Equation 4.2:

$$Info(5,4) = -\frac{5}{9}\log_2\frac{5}{9} - \frac{4}{9}\log_2\frac{4}{9}$$
$$= (0.471) + (0.519) = \mathbf{0.99}$$

TABLE 4.6

Number of Class Attributes for Various Feed Values

Feed	No. of Low (L)	No. of High (H)
0.1	1	2
0.15	1	2
0.2	2	1

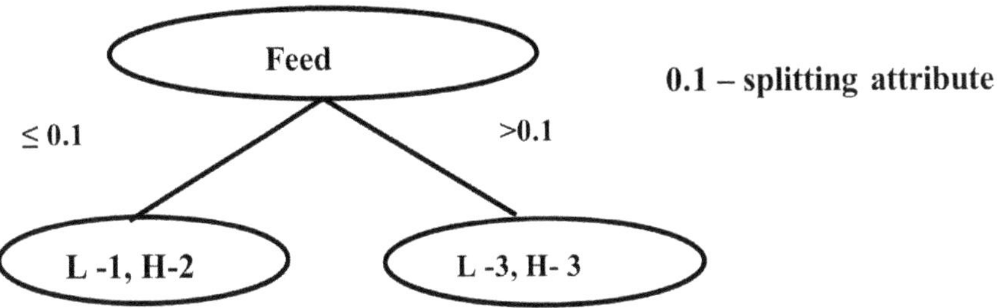

FIGURE 4.3 Decision tree for attribute feed.

The Info (5, 4) shows the overall gain for machining time.
From Equation 4.2:

$$Info(1,2) = -\frac{1}{3}log_2\frac{1}{3} - \frac{2}{3}log_2\frac{2}{3} = 0.917$$

$$Info(3,3) = -\frac{3}{6}log_2\frac{3}{6} \times 2 = 1$$

$$Combined\ Info\big((1,2),(3,3)\big) = \frac{3}{9} \times (0.917) + \frac{6}{9} \times 1 = \mathbf{0.971}$$

From Equation 4.3:

$$Gain\ of\ attribute\ Feed = (0.99) - (0.971) = \mathbf{0.026}$$

2. **Information gain for the attribute DoC**
   To calculate the IG of the DoC attribute, the number of low and high
   class values for corresponding instance numbers are noted and shown in
   Table 4.7 and Figure 4.4.

   From Equation 4.2:

$$Info(1,2) = -\frac{1}{3}log_2\frac{1}{3} - \frac{2}{3}log_2\frac{2}{3} = 0.917$$

$$Info(3,3) = -\frac{3}{6}log_2\frac{3}{6} \times 2 = 1$$

$$Combined\ Info\big((1,2),(3,3)\big) = \frac{3}{9} \times (0.917) + \frac{6}{9} \times 1 = \mathbf{0.971}$$

TABLE 4.7
Number of Class Attributes for Various Feed Values

Doc	No. of Low (L)	No. of High (H)
0.1	1	2
0.2	1	2
0.3	2	1

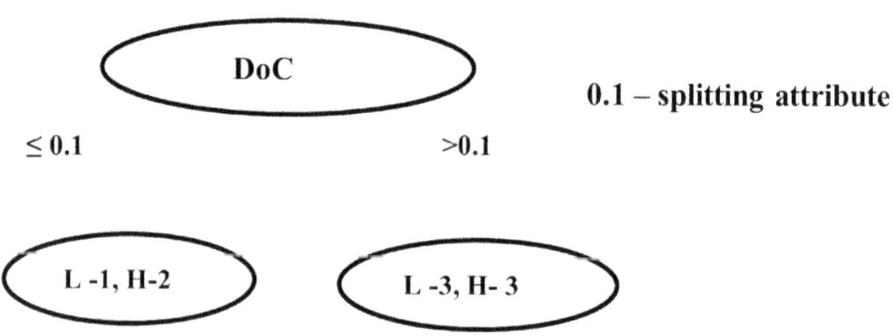

FIGURE 4.4   Decision tree for attribute DoC.

From Equation 4.3:

$$Gain\,of\,attribute\,DoC = (0.99) - (0.971) = \mathbf{0.026}$$

The values of IG of all predictor attributes are tabulated in Table 4.8. The attribute which possesses the highest IG is taken as the splitting attribute. In the previous case, the parameter speed possesses the maximum IG. The tree structure for speed is shown in Figure 4.5.

The rules derived from the root node are as follows:

if Speed ≤ 1500 rpm,
then machining time is high

Because the left side of the tree cannot be split any further, it is referred to as a "leaf node," whereas on the right side of tree, it is possible to split further; hence it is called a "growing node." The rule derived from the leaf node is "if speed is ≤ 1500 rpm, then machining time is high." This rule correctly classifies instances 1, 2 and 3 in Table 4.2. Therefore, the accuracy of the tree is 100%.

TABLE 4.8
Information Gain of Various Predictor Attributes

Sl. No.	Predictor Attributes	Information Gain
1	Speed	0.378
2	Feed	0.026
3	Doc	0.026

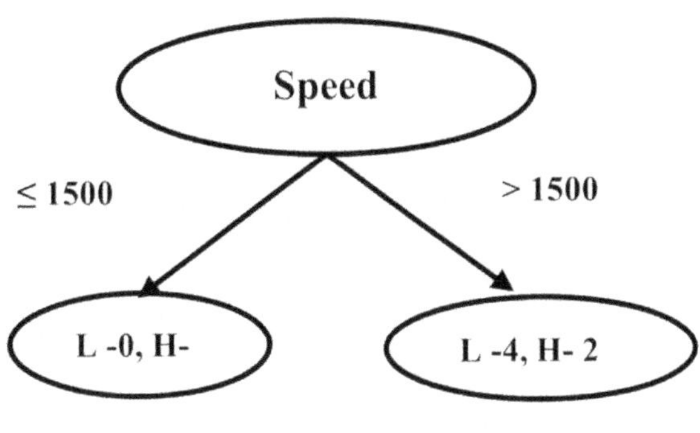

FIGURE 4.5 Tree structure of speed.

4.9 CONSTRUCTION OF SUBNODES

Instances 1, 2 and 3 are removed from Table 4.2, and remaining instances are listed in Table 4.9. For the remaining instances, once again, the IGs of both input and output variables are calculated. The tree will be processed recursively. In this case, the total number of low is 4 and the number of high is 2 for remaining instances corresponding to machining time.

$$\text{Therefore, } Info(4,2) = -\frac{4}{6} log_2 \frac{4}{6} - \frac{2}{6} log_2 \frac{2}{6} = 0.917.$$

4.9.1 IG OF INPUT VARIABLES

The calculation of the IG of input variables is discussed in the following section.

1. **IG of speed**

   The number of low and high class values for corresponding instance numbers is documented and represented in Table 4.10 and Figure 4.6 to calculate the IG of the attribute speed.

TABLE 4.9

Remaining Instances after Root Node

Instance No.	Input Variables (Predictor Attributes)			Output Variables (Class Attributes)		
	Speed (rpm)	Feed (mm/rev)	DoC (mm)	Machining Time (min)	Surface Finish	MRR
4	1750	0.1	0.2	High	Good	Low
5	1750	0.15	0.3	High	Good	Low
6	1750	0.2	0.1	Low	Bad	Low
7	2000	0.1	0.3	Low	Good	High
8	2000	0.15	0.1	Low	Good	High
9	2000	0.2	0.2	Low	Bad	Low

TABLE 4.10

Number of Class Attributes for Various Speed Values

Speed	No. of Low (L)	No. of High (H)
1750	1	2
2000	3	0

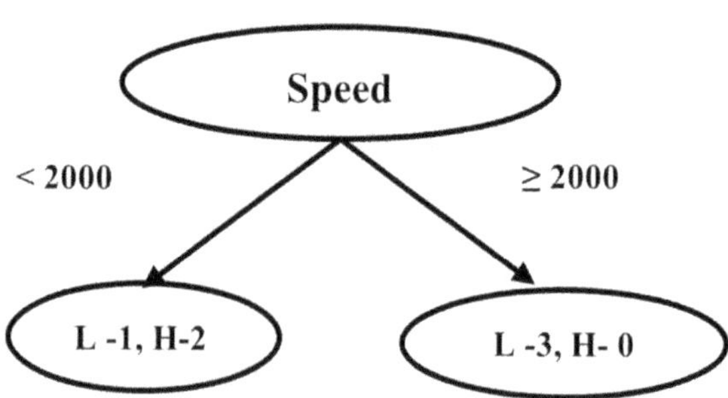

FIGURE 4.6 Tree structure of speed.

From Equation 4.2:

$$Info(1,2) = -\frac{1}{3}\log_2\frac{1}{3} - \frac{2}{3}\log_2\frac{2}{3} = 0.917$$

$$Info(3,0) = 0$$

$$Combined\ Info\big((1,2),(3,0)\big)=\frac{3}{9}\times(0.917)+\frac{3}{6}\times 0=\mathbf{0.458}$$

From Equation 4.3:

$$Gain\ of\ attribute\ Speed=(0.917)-(0.458)=\mathbf{0.458}$$

2. **IG of feed**

The number of low and high class values for corresponding instance numbers is documented and represented in Table 4.11 and Figure 4.7 to compute the IG of the attribute feed.

From Equation 4.2:

$$Info(2,2)=-\frac{4}{6}log_2\frac{4}{6}\times 2=0.917$$

TABLE 4.11

Number of Class Attributes for Various Feed Values

Feed	No. of Low (L)	No. of High (H)
0.1	1	1
0.15	1	1
0.2	2	0

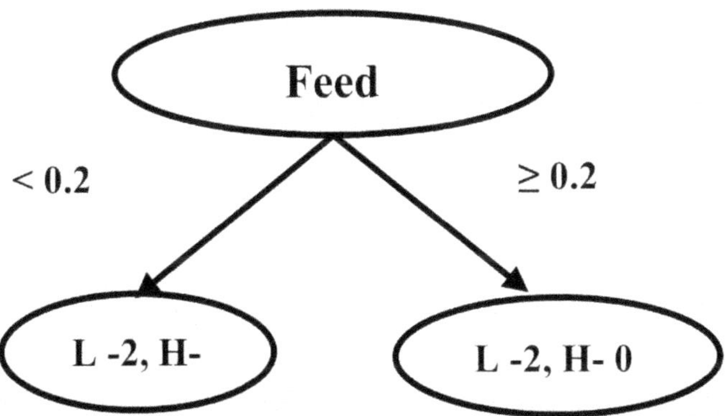

FIGURE 4.7 Tree structure of feed.

$$Info(2,0)=0$$

$$Combined\ Info((2,2),(2,0))=\frac{4}{6}\times(0.917)+\frac{2}{6}\times0=\mathbf{0.666}$$

From Equation 4.3:

$$Gain\ of\ attribute\ Feed=(0.917)-(0.666)=\mathbf{0.25}$$

3. **IG of DoC**

   The number of low and high-class values for the corresponding instance numbers are documented and represented in Table 4.12 and Figure 4.8 to compute the IG of the DoC attribute.

From Equation 4.2:

$$Info(2,2)=-\frac{4}{6}log_2\frac{4}{6}\times2=0.917$$

TABLE 4.12

Number of Class Attributes for Various DoC Values

DoC	No. of Low (L)	No. of High (H)
0.1	1	1
0.2	1	1
0.3	2	0

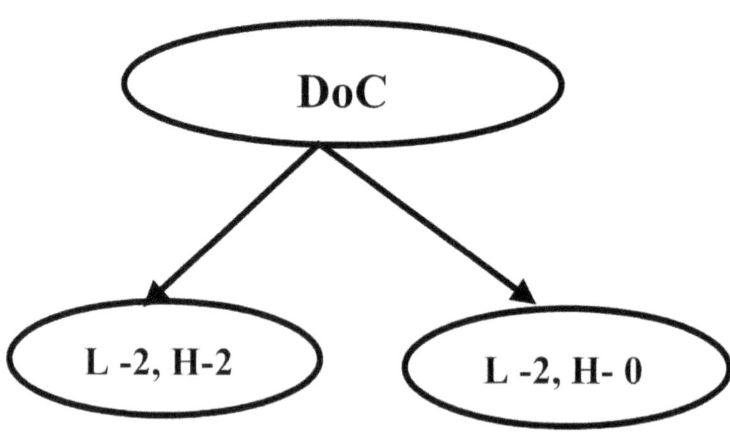

FIGURE 4.8 Tree structure of DoC.

$$Combined\ Info\big((2,2),(2,0)\big) = \frac{4}{6} \times (0.917) + \frac{2}{6} \times 0 = \mathbf{0.666}$$

From Equation 4.3:

$$Gain\ of\ attribute\ DoC = (0.917) - (0.666) = \mathbf{0.25}$$

Table 4.13 summarizes the IG values for all predictor attributes. From Table 4.13, it can be concluded that speed possesses the maximum gain when compared with feed and DoC. The tree structure of the same is shown in Figure 4.9.

The rules drawn from the preceding tree are as follows:

if speed < 2000 rpm
then machining time is high
else machining time is low

The left side of the tree correctly classifies instances 4 and 5 and incorrectly classifies instance 6. The accuracy of the tree is 66.66%. The right side of the tree correctly classifies instances 7, 8 and 9, and the accuracy is 100%. Both sides of

---

TABLE 4.13

Information Gain of Various Predictor Attributes

Sl. No.	Predictor Attributes	Information Gain
1	Speed	0.458
2	Feed	0.25
3	Doc	0.25

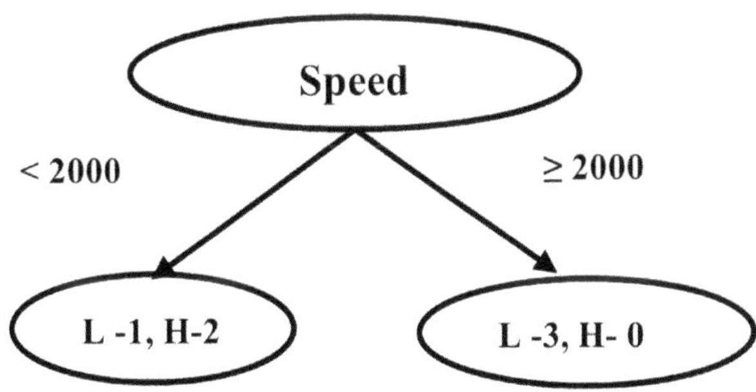

FIGURE 4.9   Tree structure for speed.

the tree cannot grow further, so the formation of the tree is stopped. The final tree structure is shown in Figure 4.10.

The decision rules formed from the previous trees are as given here:

1. if speed is ≤ 1500 rpm
   then machining time is high
2. if speed is > 1500 rpm and < 2000 rpm,
   then machining time is high
   else the machining time is low
   the overall rule which is achieved for machining time is as follows
   if speed is < 2000 rpm
   then machining time is high
   else it is low

These rules satisfy all the instances except number 6. The accuracy of the tree is 88.88%.

Observations inferred from the decision rules are listed as follows.

1. During the hard turning of OHNS steels, if feed is ≤ 0.2 mm/rev, DoC is ≤ 0.3 mm and speed is ≤ 1750 rpm, then machining time is high; else machining time is low.
2. Speed is the most important parameter for machining time as compared to feed and DoC.

These rules match the experimental results achieved during the hard turning of OHNS.

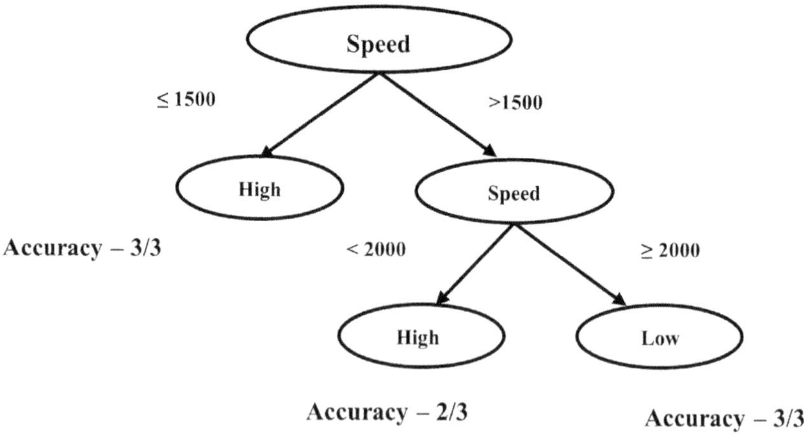

FIGURE 4.10   Decision tree for L9 orthogonal array for machining time.

4.10 DECISION TREE FOR SURFACE ROUGHNESS

Similar to machining time, data mining is constructed for surface roughness. For brevity, steps are omitted. Table 4.14 represents the various attributes of IG. From Table 4.14, it can be seen that feed and DoC have the same values of IG. The decision tree can be constructed using predictor attributes. The tree structure for feed is shown in Figure 4.11.

The decision rules which can be inferred from the previous decision tree are as follows:

> if feed is ≤ 0.15 mm/rev
> then surface roughness is good
> else 'bad'

(OR)

> if doc is ≤ 0.2 mm
> then surface roughness is good
> else bad

Except for case number 3, the aforementioned rules correctly match all instances. The accuracy of the tree is 88.88%. During the machining process, if feed is

TABLE 4.14

Information Gain of Various Predictor Attributes

Sl. No.	Predictor Attributes	Information Gain
1	Speed	0.15
2	Feed	0.456
3	Doc	0.456

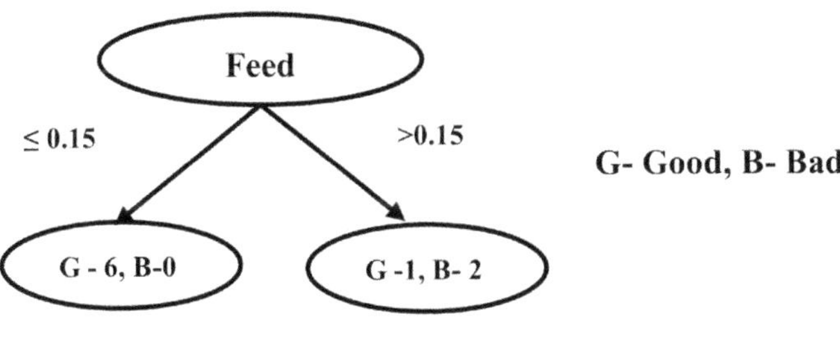

FIGURE 4.11 Decision tree for L9 orthogonal array for surface roughness.

≤ 0.15 mm/rev and DoC is ≤ 0.2 mm, then the surface roughness is good. Speed is the least significant factor when compared to feed and DoC.

4.11 DECISION TREE FOR MRR

Similar to machining time, data mining is constructed for MRR. For brevity, steps are omitted, and the final tree is shown in Figure 4.12. The IG for various attributes is shown in Table 4.15.

The decision rules which are inferred from Figure 4.12 are as follows

1. if speed is ≥ 2000 rpm
   then the MRR is high

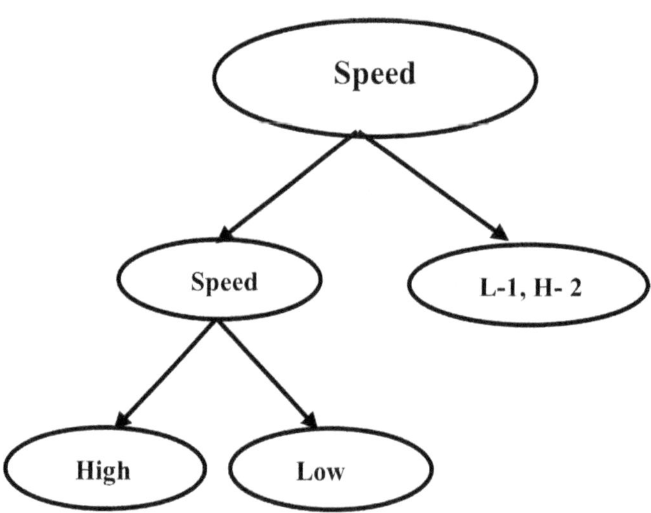

FIGURE 4.12   Decision tree for L9 orthogonal array for MRR.

TABLE 4.15
Information Gain of Various Predictor Attributes

Sl. No.	Predictor Attributes	Information Gain
1	Speed	0.082
2	Feed	0.026
3	Doc	0.026

2. if speed is < 2000 rpm and ≤ 1500 rpm,
   then MRR is high
   else the MRR is low
   the overall rule is:
   if speed is ≤ 1500 rpm and ≥ 2000 rpm
   then MRR is high
   else it is low

The inferences concluded from these rules are as follows:

1. During the hard turning of OHNS, if feed is ≤ 0.2 mm/rev and DoC is ≤ 0.3, when speed is ≤ 2000 rpm, the MRR is high.
2. Speed is the most important parameter for MRR as compared to feed and DoC.

These rules also match the experimental results.

4.12 CONCLUSION

The machining of OHNS was carried out successfully by the dry turning process using a carbide insert according to the L9 orthogonal array. The experiments were carried out with input parameters like (1) speed, (2) feed and (3) DoC. The output parameters measured were (1) MRR, (2) surface roughness and (3) machining time. In the present work, the C 4.5 DT algorithm is applied for the output of OHNS machining parameters, and the rules are framed based on the attribute. In the case of speed, if it is less than 2000 rpm, the machining time was higher. When surface finish is considered, if the feed is less than or equal to 0.15, the surface roughness is good, and in the case of DoC for surface roughness, it must be greater than or equal to 0.2 mm. For maximum MRR, the machining speed must be greater than or equal to 2000 rpm. The rules formed based on the attributes matched with the results of the experiment.

REFERENCES

Ahmed, Mohamed Ahmed, Ahmet Rizaner, and Hakan Ulusoy Ali. 2018. "A Novel Decision Tree Classification Based on Post-Pruning with Bayes Minimum Risk." *PLoS One* 13 (4): 1–12. https://doi.org/10.1371/journal.pone.0194168.

Balachandhar, R., R. Balasundaram, and M. Ravichandran. 2021. "Analysis of Surface Roughness of Rock Dust Reinforced AA6061-Mg Matrix Composite in Turning." *Journal of Magnesium and Alloys* 9 (15): 1669–1676. https://doi.org/10.1016/j.jma.2021.03.035.

Govindan, Kannan, R. Balasundaram, N. Baskar, and P. Asokan. 2017. "A Hybrid Approach for Minimizing Makespan in Permutation Flowshop Scheduling." *Journal of Systems Science and Systems Engineering* 26 (1): 50–76. https://doi.org/10.1007/s11518-016-5297-1.

Huber, Norbert, Surya R. Kalidindi, Benjamin Klusemann, and Christian J. Cyron. 2020. "Editorial: Machine Learning and Data Mining in Materials Science." *Frontiers in Materials* 7 (February): 1–3. https://doi.org/10.3389/fmats.2020.00051.

James Dhilip, Jafrey Daniel, J. Jeevan, D. Arulkirubakaran, and M. Ramesh. 2020. "Investigation and Optimization of Parameters for Hard Turning of OHNS Steel." *Materials and Manufacturing Processes* 35 (10): 1113–1119. https://doi.org/10.1080/10426914.2020.1765254.

Katz, Gilad, Asaf Shabtai, Lior Rokach, and Nir Ofek. 2014. "Confdtree: A Statistical Method for Improving Decision Trees." *Journal of Computer Science and Technology* 29 (3): 392–407. https://doi.org/10.1007/s11390-014-1438-5.

Koenig, W., R. Komanduri, H. K. Toenshoff, and G. Ackershott. 1984. "Machining of Hard Materials." *CIRP Annals* 33: 417–427. Technische Rundschau. https://doi.org/10.1016/s0007-8506(16)30164-0.

Singh, Dilbag, and P. Venkateswara Rao. 2007. "A Surface Roughness Prediction Model for Hard Turning Process." *International Journal of Advanced Manufacturing Technology* 32 (11–12): 1115–1124. https://doi.org/10.1007/s00170-006-0429-2.

Tonshoff, H. K., C. Arendt, and R. Ben Amor. 2000. "Cutting of Hardened Steel." *CIRP Annals—Manufacturing Technology* 49 (2): 547–566. https://doi.org/10.1016/S0007-8506(07)63455-6.

Vereschaka, Alexey, Vladimir Tabakov, Sergey Grigoriev, Nikolay Sitnikov, Gaik Oganyan, Nikolay Andreev, and Filipp Milovich. 2019. "Investigation of Wear Dynamics for Cutting Tools with Multilayer Composite Nanostructured Coatings in Turning Constructional Steel." *Wear* 420–421 (November 2018): 17–37. https://doi.org/10.1016/j.wear.2018.12.033.

5 Effect of Moisture Content on Geometrically Nonlinear Transient Vibrations of Laminated Composite Plates

Srinivas Prabhu M, Ashok M H, Vishwanath Khadakbhavi, R. M. Galagali, J. Shivakumar, Nithin Kumar, and Rashmi P Shetty

5.1 INTRODUCTION

Laminated composite structures are widely used in the structural fields of aerospace, automotive, biomedical and civil structures. Composite materials have high strength to weight ratio, high stiffness and high flexural rigidity, improved toughness, corrosion resistance and resistance to oxidation. These properties are superior compared to conventional ceramics and alloys. In space engineering applications, composites are expected to perform very efficiently under the influences of moisture content, higher temperatures and chemical environments. Challenging and hazardous conditions may subject composites to large deformations and bending behaviors. Therefore, it is important to carry out a careful investigation on the geometrically nonlinear behavior of composites in such extraordinary environments.

5.2 LITERATURE SURVEY

Many researchers have performed various analyses of composites under hygrothermal environments and presented their work. M K Rath and S K Sahu [1] have studied the vibrations of woven fiber-laminated composite plates under hygrothermal environments. B P Patel and coauthors [2] have investigated the effect of hygrothermal environments on the structural behavior of thick composite laminates with higher-order theory. S K Singh and A Chakraborti [3] have performed hygrothermal analysis of laminated composite plates using efficient higher-order

DOI: 10.1201/9781003397465-5

"

shear deformation theory. Xiangyang Li and co researchers [4] have emphasized the buckling and vibro-acoustic response of clamped composite laminated plates in thermal environments. L S Ramachandra and Sarat Kumar Panda [5] have studied the dynamic instability of composite plates subjected to non-uniform in-plane loads. H R Ovesy and J Fazilati [6] have focused on parametric instability analysis of moderately thick functionally graded cylindrical panels using finite strip methods. The buckling of symmetrically laminated rectangular plates under parabolic edge compressions has been investigated by Yuhua Tang and Xinwei Wang [7]. M. Rafiee and coauthors [8] have conducted research on the non-linear dynamic stability of piezoelectric functionally graded carbon nanotube-reinforced composite plates with initial geometric imperfection. Aman Garg and H D Chalak [9] have presented a review on the analysis of laminated and sandwich structures under hygrothermal conditions. Hygrothermal effects on the linear and nonlinear analyses of composite plates have been carried out by Sen Yung Lee and coauthors [10]. The geometrically nonlinear static analysis of laminated smart composite plates integrated with patches of active fiber composites (AFCs) and piezoelectric fiber-reinforced composite (PFRC) materials acting as distributed actuators has been compared using generalized energy-based finite-element models [11]. The work focuses on geometrically nonlinear transient analysis of laminated smart composite plates integrated with the patches of active fiber composites using active constrained layer damping (ACLD) as distributed actuators. The analysis has been carried out using a generalized energy based finite-element model [12]. A thin laminated composite shallow shell as smart structure with AFC material's ACLD treatment was analyzed for geometrically nonlinear transient vibrations. The AFC material was used to make the constraining layer of the ACLD treatment [13]. The work was carried out on a comparative investigation for the performance of an active constrained layer damping treatment using PFRC and AFC patch material for controlling the nonlinear transient vibrations of a laminated composite plate [14]. In one paper, a comparative investigation of the performance of an active constrained layer damping treatment using PFRC and AFC patch material for controlling the nonlinear transient vibrations of a laminated composite shallow shell was carried out [15]. In another paper, analytical solutions were obtained for simply supported plates under static electrical and mechanical loads. The results were validated with existing 3D elasticity solutions and compared with other plate theory solutions [16].

The electro-mechanical behavior of laminated composite plate integrated with PFRC actuator at the top surface of the plate using the trigonometric zigzag theory (TZZT) were carried out. The responses in the form of deflection, normal stress and transverse shear stress of the smart structure were examined for various electro-mechanical loading and span-thickness ratios. The actuating effects were much more pronounced in thick laminates. The PFRC layer had much more controlling authority over thick laminates [17]. The performance of vertically reinforced 1–3 piezoelectric composite material was performed as a constraining layer for active constrained layer damping of sandwich beams. A finite-element model, considering both in-plane and out-of-plane actuation of the constraining layer of active constrained layer damping, has been developed for analyzing the active damping of sandwich beams integrated with the patches of the active constrained layer damping treatment. The analysis revealed that vertically reinforced 1–3 piezo composites, which are popularly used as sensor materials, can be used as distributed actuators of smart sandwich beams [18].

The analysis work in [19] presents the development of a novel non-polynomial electromechanical shell theory for the analysis of composite piezoelectric shell structures. The 2D shell theory is utilized to explore novel shear sensors made of low-symmetry materials with simple patch geometries. Parametric studies are subsequently performed to analyze the effect of geometric parameters such as thickness, radius of curvature of the shell, boundary conditions and thickness of the piezoelectric layer on the sensor response of piezoelectric shell structures. It is observed that the response of these sensors is highly dependent on the boundary conditions and placement and thickness of the piezoelectric layer [19]. The electromechanical responses of smart laminated composite plates with piezoelectric materials are derived using a two-dimensional displacement-based non-polynomial higher-order shear deformation theory. The responses obtained in the form of deflection and stresses are compared with three-dimensional solutions and also with different polynomial and non-polynomial based higher-order theories in the literature. The transverse shear stresses are obtained using 3D equilibrium equations of elasticity to enhance the accuracy of the present results. Various examples are numerically solved to establish the efficiency of the present model [20].

Ghugal and Kulkarni [21] presented an equivalent single-layer shear deformation theory for evaluation of displacements and stresses of cross-ply laminated plates subjected to uniformly distributed nonlinear thermo-mechanical loads. The numerical results were compared with those of classical, first-order and higher-order shear deformation plate theories. Later they [22] studied the flexural response of symmetric cross-ply laminated plates subjected to uniformly distributed linear and non-linear thermo-mechanical loads. The results were compared with those of classical plate theory, first-order shear deformation theory and higher-order shear deformation theory. Kerur and Ghosh [23] worked on a finite-element formulation for the geometrically non-linear bending behavior of smart structures integrated with a piezoelectric fiber-reinforced composite (AFC/MFC) layer acting as distributed actuator. A wide variety of numerical examples considering cross-ply laminated substrates subjected to combined mechanical, hygrothermal and electrical loads were presented. The effect of piezoelectric fiber orientation in an actuator to counteract non-linear deflections was also analyzed. In the dynamic analysis, the active vibration control of piezoelectric laminated beams under thermal load are presented. The designed model controls the tip deflection of composite and sandwich cantilever beams and midpoint deflection of clamped-clamped beams [24]. This paper presented a steady-state analytical solutions for the coupled thermo-electro-elastic forced vibrations of piezoelectric laminated beams. The interactions between the thermal and electric factors were obtained analytically. The convergence of the present solutions was first verified in the numerical section, followed by the FEM results used to validate the achieved solutions. From the sample numerical calculations, it is seen that the two-dimensional temperature field presents different distributions for different electric conditions [25].

The static and dynamic responses of piezoelectric beams were investigated using the developed NS-RPIM–based beam model at different hygrothermal conditions. The effects of the damping ratios and the hygrothermal parameters on the generalized displacements and natural frequencies of a clamped energy harvester with different combinations of inside holes were comprehensively discussed. Additionally, the high accuracy and excellent convergence rate of the proposed method were also validated by using a piezoelectric cantilever beam with HTEM coupling. The presented NS-RPIM–based two-dimensional beam model can be used for future studies of coupled multi-physical problems as well as investigations of complex piezoelectric sensors and actuators in a hygrothermal environment [26]. In another paper, isogeometric analysis (IGA) integrated with a simple first-order shear deformation theory (S-FSDT) with only four variables per node was extended to study the geometrically nonlinear natural frequency, nonlinear static bending and transient dynamic response of two types of smart piezoelectric functionally graded plates (PFGPs) in a thermal environment. To investigate the nonlinear mechanical behavior of PFGPs in a thermal environment, closer to the real situation, an augmented piezoelectric constitutive equation considering the thermal effects of piezoelectric stress and dielectric constants was used. The accuracy and effectiveness of the developed approach are demonstrated through numerical experiments with comparisons. The authors additionally investigated the variation of temperature distribution on the geometrically nonlinear response of PFGPs suffering thermo-electro-mechanical load [27].

5.3 FINITE-ELEMENT MODELING

Figure 5.1 represents N layers of orthotropic layers of a smart rectangular laminated composite plate. On the top surface of this plate, two rectangular patches with ACLD treatment are placed. The piezoelectric fiber-reinforced composite/active fiber composite is used in the constraining layer of the ACLD treatment. PFRC/AFC has a layer thickness of h_p, and the viscoelastic material of the ACLD has a thickness of h_v. The dimensions of the plate are $a/2$ and $b/2$ as length and width, respectively. Figure 5.1 shows that exactly at the middle of the substrate plate, the ACLD treatment patches are placed.

The mid-plane of the substrate plate is taken as a reference plane, and the origin of the laminate co-ordinate system (x, y, z) is chosen on the reference plane. The boundaries of the total plate coincide with the line $x = 0, a$ and $y = 0, b$. Also, h_{k+1} and h_k represent the thickness of the substrate, and k represents the number of layers. The fiber orientation angle is denoted by θ with regard to the laminate coordinate system. The orientations of the fiber in the PFRC/AFC layer are given by j. The alignment of the fibers in the substrate is in the longitudinal direction.

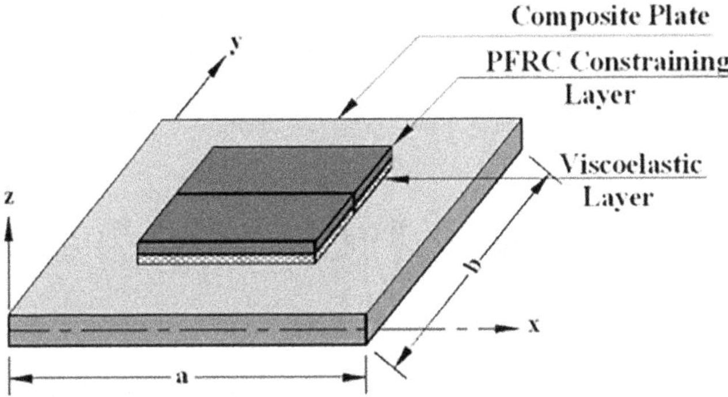

FIGURE 5.1 Laminated composite plate under ACLD treatment with PFRC/AFC patches.

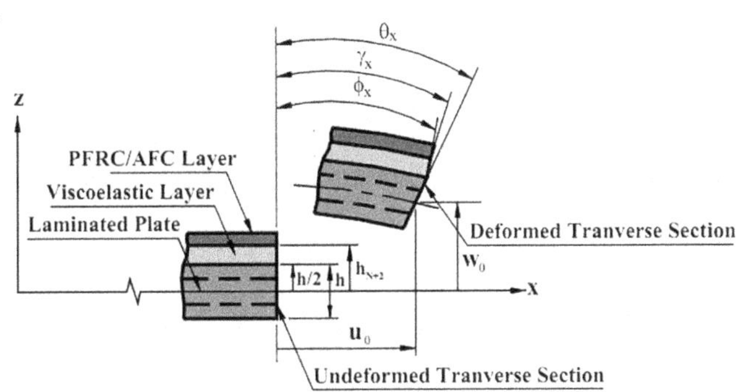

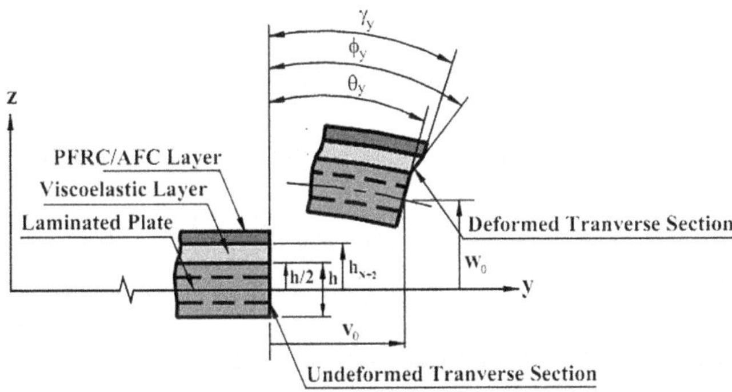

FIGURE 5.2 Deformation of any transverse cross-section of the laminated plate integrated with the ACLD treatment, which is parallel to the: (a) yz plane (b) xz plane.

In this experiment, first-order shear deformation theory is applied for axial displacements and modeling the kinematics of deformations of the plate coupled with the ACLD treatment. Figures 5.1 and 5.2 show the deformation. At any point in the layer, the displacement u and v is given along the x and y directions and mathematically illustrated by

$$u(x,y,z,t) = u_0(x,y,t) + \left[z - \left\langle z - \frac{h}{2}\right\rangle\right]\theta_x(x,y,t) + \left[\left\langle z - \frac{h}{2}\right\rangle - \left\langle z - h_{N+2}\right\rangle\right]\phi_x$$
$$(x,y,t) + \left\langle z - h_{N+2}\right\rangle\gamma_x(x,y,t) \tag{5.1}$$

$$v(x,y,z,t) = v_0(x,y,t) + \left[z - \left\langle z - \frac{h}{2}\right\rangle\right]\theta_y(x,y,t) + \left[\left\langle z - \frac{h}{2}\right\rangle - \left\langle z - h_{N+2}\right\rangle\right]\phi_y$$
$$(x,y,t) + \left\langle z - h_{N+2}\right\rangle\gamma_y(x,y,t) \tag{5.2}$$

$$w(x,y,z,t) = w_0(x,y,t) + \left[z - \left\langle z - \frac{h}{2}\right\rangle\right]\theta_z(x,y,t) + \left[\left\langle z - \frac{h}{2}\right\rangle - \left\langle z - h_{N+2}\right\rangle\right]\phi_z$$
$$(x,y,t) + \left\langle z - h_{N+2}\right\rangle\gamma_z(x,y,t) \tag{5.3}$$

In the equations, $\langle\ \rangle$ illustrates the singularity functions, which satisfy the continuity conditions between any two portions. The total potential energy T_P and the kinetic energy T_K can be written as:

$$T_P = \frac{1}{2}\left[\sum_{k=1}^{N+2}\int_{\Omega_k}(\{\varepsilon b^k\}^T\{\sigma b^k\} + \{\varepsilon_{sh}^k\}^T\{\sigma_{sh}^k\})d\Omega - \int_{\Omega_{N+2}} E_i D_i dv\right] \tag{5.4}$$
$$- \int_A pwdA$$

$$T_k = \frac{1}{2}\left[\sum_{k=1}^{N+2}\int_{\Omega_k}\rho^k\left(\dot{u}^2 + \dot{v}^2 + \dot{w}^2\right)d\Omega\right] \tag{5.5}$$

where **p** is the externally applied transversely distributed load acting over a surface area A, and Ωk and $k\rho$ represent the volume and mass density of the k layer. The subscript x is used for AFC and z for PFRC material in the study. The value of k is 1 to N for substrate, $N + 1$ for the viscoelastic layer, and $N + 2$ for the AFC layer, and the dot (.) over the variables denotes the derivative of the respective variable with respect to time.

The connective relations for the substrate can be given by

$$\{\sigma_b^k\} = \left[C_b^k\right]\{\varepsilon_b^k\} \text{ and } \{\sigma_{sh}^k\} = \left[C_{sh}^k\right]\{\varepsilon_{sh}^k\} \quad k = 1,2,3,......N \tag{5.6}$$

To include hygro-thermal conditions, the relations are changed to

$$\mathbf{T(a,b,z,t)} = \mathbf{T(a,b,t)} \text{ and } \mathbf{C(a,b,z,t)} = \mathbf{C(a,b,t)} \tag{5.7}$$

The electric field $\mathbf{E_i}$ is applied. Isotropic behavior of the material is assumed. The relation with a time domain analysis is given by:

$$\{\sigma_{sh}\}^{N+1} = \int_0^t G(t-\tau)\frac{\partial\{\varepsilon_{sh}\}_v}{\partial\tau}\partial\tau \tag{5.8}$$

where $\mathbf{G(t)}$ is the relaxation function. The hygro-thermal effect for the total potential energy $\mathbf{T_P^e}$ is given by

$$\mathbf{T_P^e} = \frac{1}{2}[\{\mathbf{d_t^e}\}^\mathbf{T}[\mathbf{K_{tt}^e}]\{\mathbf{d_t^e}\} + \{\mathbf{d_t^e}\}^\mathbf{T}[\mathbf{K_{tr}^e}]\{\mathbf{d_t^e}\} + \{\mathbf{d_r^e}\}^\mathbf{T}[\mathbf{K_{tr}^e}]\{\mathbf{d_t^e}\} + \{\mathbf{d_t^e}\}^\mathbf{T}[\mathbf{K_{rr}^e}]\{\mathbf{d_t^e}\}$$

$$+ \{\mathbf{d_t^e}\}^\mathbf{T}[\mathbf{K_{tsv}^e}]\int_0^t G(t-\tau)\frac{\partial}{\partial\tau}\{\mathbf{d_t^e}\}\partial\tau + \{\mathbf{d_t^e}\}^\mathbf{T}[\mathbf{K_{trsv}^e}]\int_0^t G(t-\tau)\frac{\partial}{\partial\tau}\{\mathbf{d_r^e}\}\partial\tau$$

$$+ \{\mathbf{d_r^e}\}^\mathbf{T}[\mathbf{K_{trsv}^e}]\int_0^t G(t-\tau)\frac{\partial}{\partial\tau}\{\mathbf{d_t^e}\}\partial\tau + \{\mathbf{d_r^e}\}^\mathbf{T}[\mathbf{K_{rrsv}^e}]\int_0^t G(t-\tau)\frac{\partial}{\partial\tau}\{\mathbf{d_r^e}\}\partial\tau$$

$$- \{\mathbf{d_t^e}\}\{\mathbf{F_{tpn}^e}\}V - 2\{\mathbf{d_t^e}\}\{\mathbf{F_{tbp/a}^e}\} - 2\{\mathbf{d_r^e}\}\{\mathbf{F_{rbp/a}^e}\} - \{\mathbf{d t^e}\}^\mathbf{T}[\mathbf{F_{tt}^e}] - \{\mathbf{d_r^e}\}^\mathbf{T}[\mathbf{F_{rt}^e}]$$

$$- \frac{\varepsilon_{33}V^2}{\mathbf{d_p^2}} - 2\{\mathbf{d_t^e}\}\{\mathbf{F^e}\}] \tag{5.9}$$

The kinetic energy $\mathbf{T_k^e}$ of the element is:

$$T_k^e = \frac{1}{2}\{\dot{d}_t^e\}[M^e]\{\dot{d}_t^e\} \tag{5.10}$$

in which

$$[\mathbf{M^e}] = \int_0^{a_e}\int_0^{b_e}\bar{\mathbf{m}}\{\mathbf{N_t}\}^\mathbf{T}[\mathbf{N_t}]\mathbf{dxdy} \text{ and } \bar{\mathbf{m}} = \sum_{k=1}^N \rho^k(\mathbf{h_{k+1}} - \mathbf{h_k}) + \rho^{N+1}(\mathbf{h_v}) + \rho^{N+2}(\mathbf{h_p})$$

The coupled electro-elastic open loop behavior of the laminated composite plate integrated with the patches of ACLD treatment, in which the GHM method is used to model viscoelastic material, is as follows

$$\left[M^*\right]\left\{\ddot{X}\right\}+\left[C^*\right]\left\{\dot{X}\right\}+\left[K^*\right]\left\{X\right\}=\left\{F^*\right\}+\left\{F_p^*\right\}V+\left\{F_t\right\} \quad (5.11)$$

5.3.1 Closed-Loop Model

Using a closed-loop model through the simple velocity feedback control law to apply control voltage to activate ACLD patches, we get the equation of motion governing the closed-loop dynamics of substrate plates activated by the patches of ACLD treatment as follows

$$V^j = -K_d^j \dot{w} = -K_d^j\left[U^j\right]\left\{\dot{X}\right\} \quad (5.12)$$

where $[C_d^*]=[C^*]+\sum_{j=1}^{m}K_d^j\{F_p^*\}[U^j]$ is the active damping matrix.

5.4 NUMERICAL RESULTS UNDER HYGRO-THERMO-MECHANICAL LOADING

In order to assess the performance of the ACLD patches made of PFRC/AFC material in controlling the nonlinear vibrations of laminated composite plates,

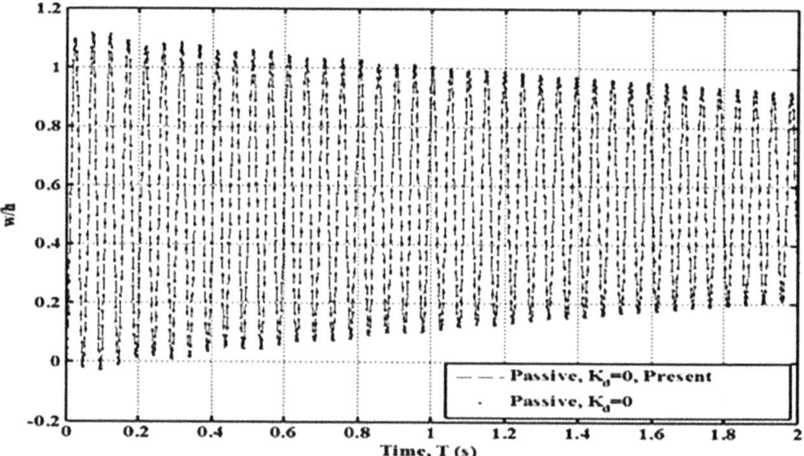

FIGURE 5.3 Comparison of present model with [28] for nonlinear dynamic response of a simply supported (SS1) symmetric cross-ply $(0^o / 90^o / 0^o)$ square substrate plate undergoing active constrained layer damping (for $a / h = 300$, $Q = 100$) using PFRC under passive mode ($K_d = 0$).

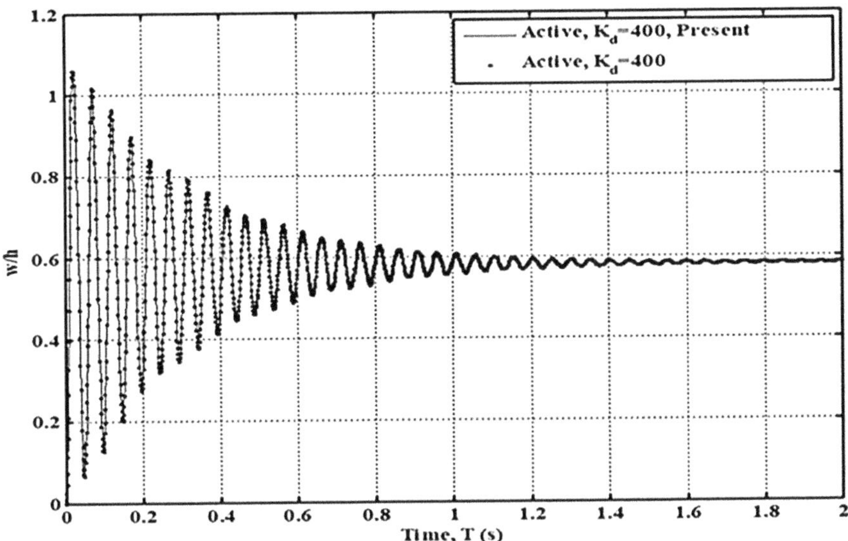

FIGURE 5.4 Comparison of present model with [28] for nonlinear dynamic response of a simply supported symmetric cross-ply $(0^o \, / \, 90^o \, / \, 0^o)$ square substrate plate undergoing active constrained layer damping (for $\mathbf{a} \, / \, \mathbf{h} = \mathbf{300}$, $\mathbf{Q} = 100$) using PFRC under active mode $(\mathbf{K_d} \neq \mathbf{0})$.

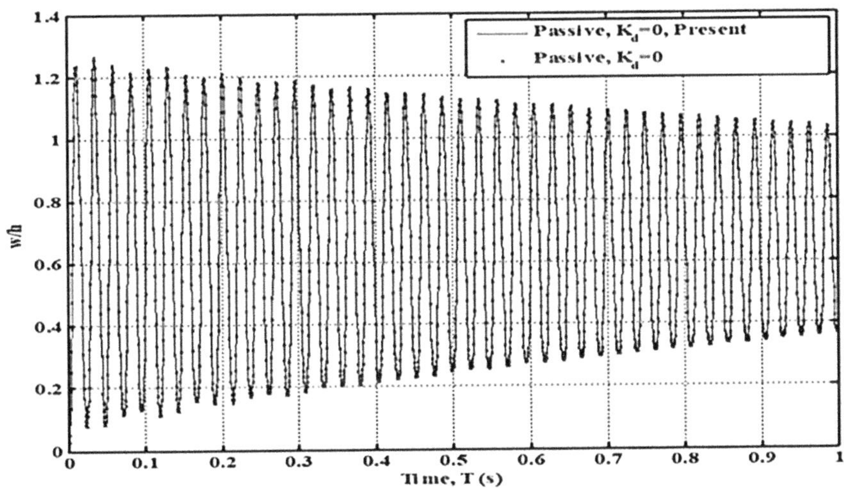

FIGURE 5.5 Comparison of present model with [29] for nonlinear dynamic response of a simply supported symmetric cross-ply $(0^o \, / \, 90^o \, / \, 0^o)$ square substrate plate undergoing active constrained layer damping (for $\mathbf{a} \, / \, \mathbf{h} = \mathbf{200}$, load = 500 N/m²) using AFC material under passive mode $(\mathbf{K_d} = \mathbf{0})$.

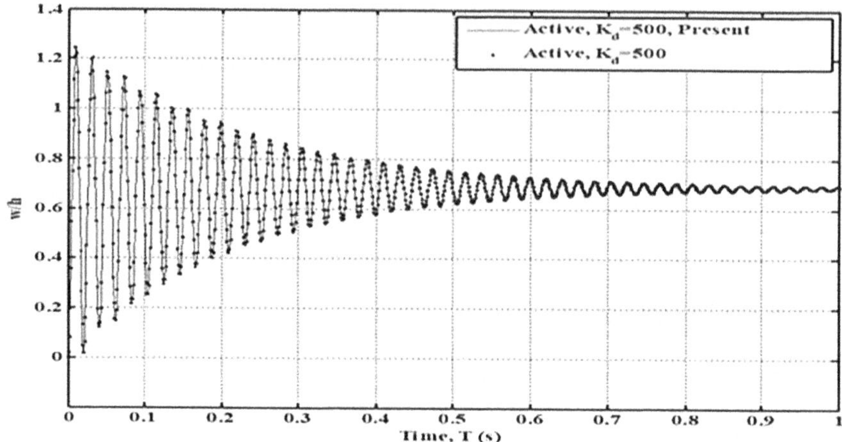

FIGURE 5.6 Comparison of present model with [29] for nonlinear dynamic response of a simply supported symmetric cross-ply $(0^o \, / \, 90^o \, / \, 0^o)$ square substrate plate undergoing active constrained layer damping (for $\mathbf{a \, / \, h = 200}$, load = 500 N/m²) using AFC material under active mode ($\mathbf{K_d \neq 0}$).

numerical results are computed using the finite-element model derived in the previous section neglecting hygro-thermal loading. The comparison of the present model with the existing model [29] is shown in Figures 5.3, 5.4, 5.5, and 5.6. Both cross-ply and angle-ply thin square substrates integrated with two patches of ACLD treatment (Figure 5.1) are considered for evaluating the numerical results. The same elastic and piezoelectric properties of the orthotropic layers of substrate plates and the PFRC patches considered for evaluating the numerical results in the previous sections are used to compute the numerical results shown here. The thicknesses of the PFRC/AFC patch, the viscoelastic patch and the laminated plates are modeled as 250 μm, 50.8 μm and 3 mm, respectively. Also, the orthotropic layers of the substrate plate are of equal thickness. Unless otherwise mentioned, the piezoelectric fiber orientation angle ψ in the PFRC/AFC patch is considered 0°. Considering a single-term GHM expression, the values of α, $\hat{\xi}$ and $\hat{\omega}$ are used as 11.42, 1.0261e5 and 20, respectively. The shear modulus $\mathbf{G^\infty}$ and the density of the viscoelastic material ρ_v are 1.822e6 Pa and 1104 kg/m³, respectively. The mechanical load $\mathbf{p}$ acting upward is assumed to be uniformly distributed, while the aspect ratio $\mathbf{a \, / \, h}$ is considered to be 300. The boundary conditions used for evaluating the numerical results are considered as follows:

$$\text{Simply supported}: \mathbf{v_0 = w_0 = \theta_y = \phi_y = \gamma_y = 0 \ at \ x = 0, a}$$

$$\mathbf{u_0 = w_0 = \theta_x = \phi_x = \gamma_x = 0 \ at \ y = 0, b}$$

The results are obtained with different moisture content conditions. The elastic properties are the same as for PFRC/AFC substrate plates, which are used for evaluating the results. The coefficient of thermal expansion constants are taken as $\alpha_1 = 0.02 \times 10^{-6}$m/m/°C and $\alpha_2 = 22.5 \times 10^{-6}$m/m/°C, and the moisture constants as $\beta_1 = 0$ and $\beta_2 = 0.44$.

To test the present finite-element model under the influence of moisture, the negligibly small thickness of the patches of ACLD treatment is considered, and the mid-plane center of the substrate plate is computed by the present model using the properties given in Table 5.1 and is compared with existing results for an identical composite plate without integration with the ACLD patches. It may be observed from Figure 5.7 that in all cases, the results are in good agreement.

$$G_{13} = G_{12}, G_{23} = 0.5G_{12}, \upsilon_{11} = 0.3, \beta_{11} = 0, \beta_{22} = 0.44, \rho = 1600 \, \text{kg} / \text{m}^3 \quad (5.13)$$

The nonlinear dynamic behavior and active control of laminated composite plates is studied under a hygro-thermal environment combined with a mechanical load with appropriate boundary conditions given in the preceding section.

Similarly, varying moisture conditions for symmetric cross-ply $(0^o / 90^o / 0^o)$, for anti-symmetric cross-ply $(0^o / 90^o / 0^o / 90^o)$ and for anti-symmetric angle ply $(-45^o / 45^o / -45^o / 45^o)$ are represented in Figures 5.8 through 5.10, respectively, for PFRC material.

Similarly, for ACLD treatment made of AFC material, the performance under the hygro-thermal environment is demonstrated in Figures 5.11 to 5.13. The figures show the performance of the AFC material in varying moisture concentration conditions for symmetric cross-ply $(0^o / 90^o / 0^o)$, anti-symmetric cross-ply $(0^o / 90^o / 0^o / 90^o)$ and anti-symmetric angle-ply $(-45^o / 45^o / -45^o / 45^o)$, respectively.

TABLE 5.1

Elastic Moduli of Graphite/Epoxy Lamina at Different Moisture Concentrations

Elastic Moduli (GPa)	Moisture Concentration, C (%)						
	0	0.25	0.50	0.75	1.0	1.25	1.50
E_{11}	130	130	130	130	130	130	130
E_{12}	9.5	9.25	8.75	8.5	8.5	8.5	8.5
G_{12}	6.0	6.0	6.0	6.0	6.0	6.0	6.0

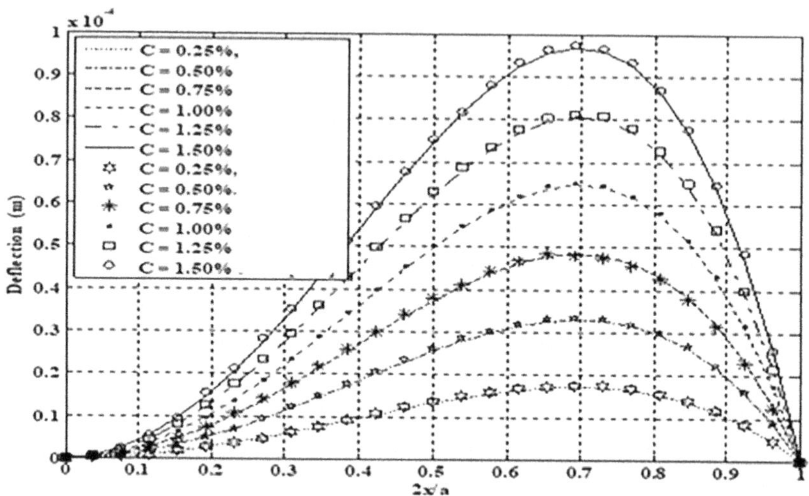

FIGURE 5.7 Comparison of deflection along x axis at different moisture concentrations [16].

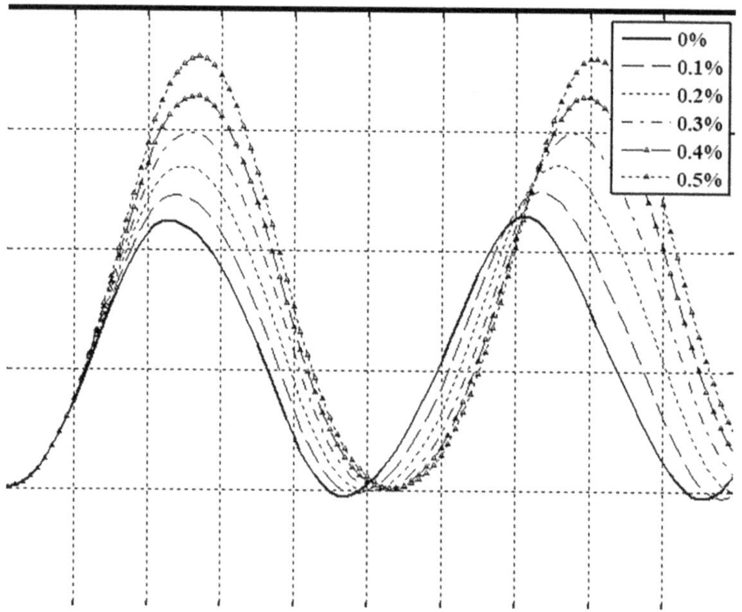

FIGURE 5.8 Nonlinear dynamic response of a simply supported anti symmetric cross-ply $(0^o / 90^o / 0^o)$ square substrate plate undergoing active constrained layer damping (for $\mathbf{a} / \mathbf{h} = \mathbf{300}$, $\mathbf{Q} = 100$) using PFRC for various moisture conditions: (a) under passive mode ($\mathbf{K_d} = \mathbf{0}$); (b) under active mode ($\mathbf{K_d} = \mathbf{75}$).

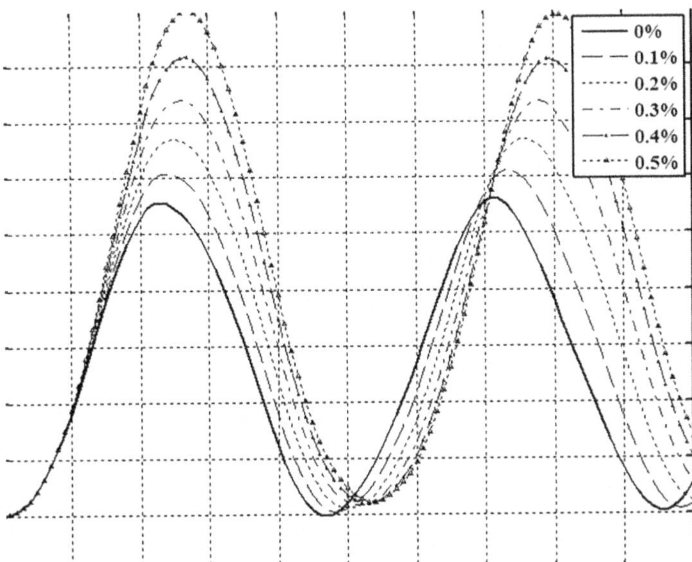

FIGURE 5.8 (Continued)

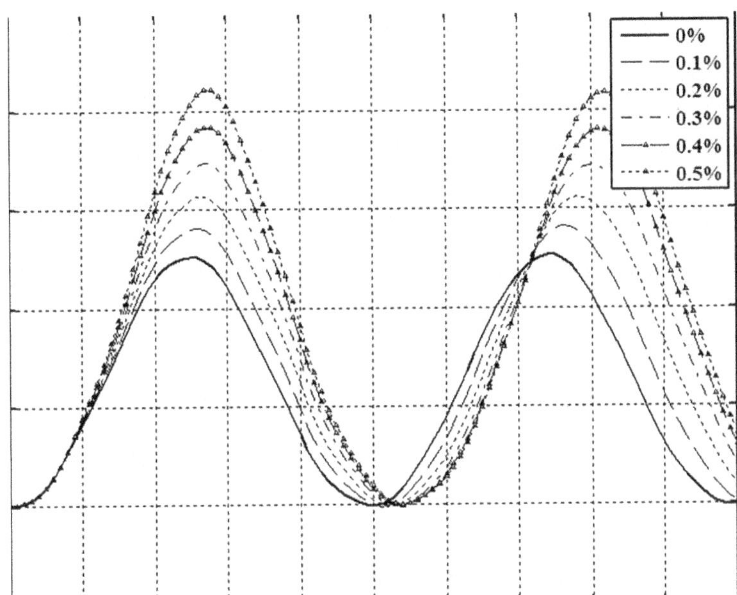

FIGURE 5.9 Nonlinear dynamic response of a simply supported anti-symmetric cross-ply $(0^o / 90^o / 0^o / 90^o)$ square substrate plate undergoing active constrained layer damping (for $a / h = 300$, $Q = 100$) using PFRC for various moisture conditions: a) under passive mode ($K_d = 0$); b) under active mode ($K_d = 75$).

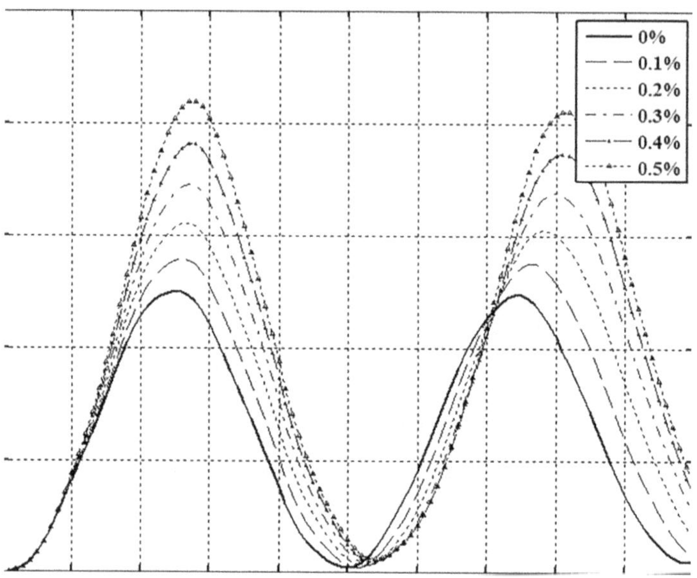

FIGURE 5.9 (Continued)

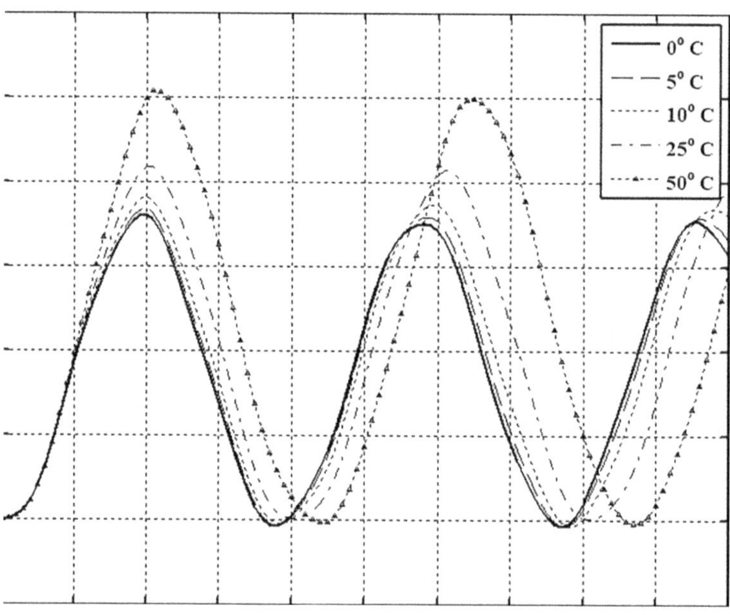

FIGURE 5.10 Nonlinear dynamic response of a simply supported anti-symmetric angle-ply $(-45^o / 45^o / -45^o / 45^o)$ square substrate plate undergoing active constrained layer damping (for $a / h = 300$, $Q = 100$) using PFRC for various moisture conditions: a) under passive mode; ($K_d = 0$) b) under active mode ($K_d = 75$).

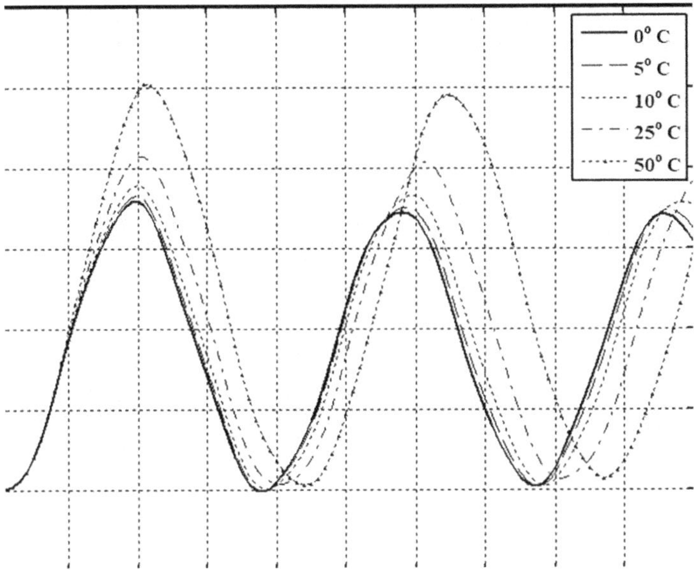

FIGURE 5.10   (Continued)

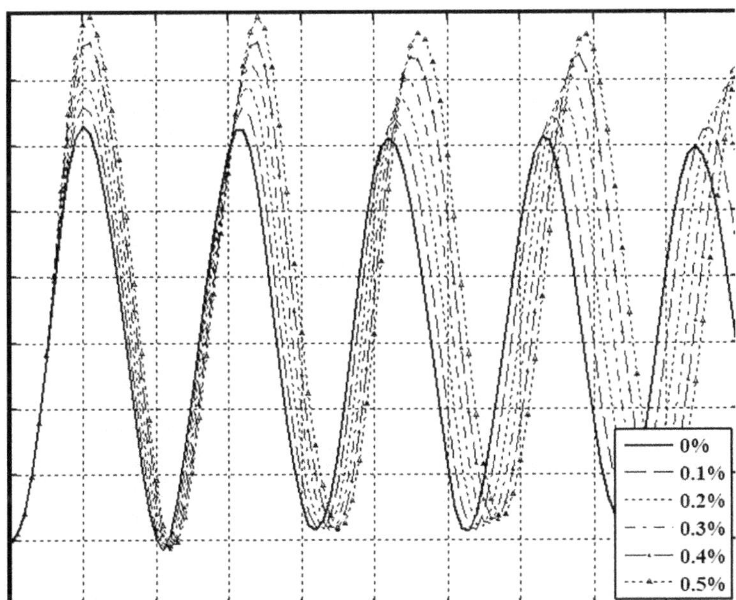

FIGURE 5.11   Nonlinear dynamic response of a simply supported anti-symmetric cross-ply $(0^o / 90^o / 0^o)$ square substrate plate undergoing active constrained layer damping (for $a / h = 300$, $Q = 100$) using AFC for various moisture conditions: a) under passive mode ($K_d = 0$); b) under active mode ($K_d = 100$).

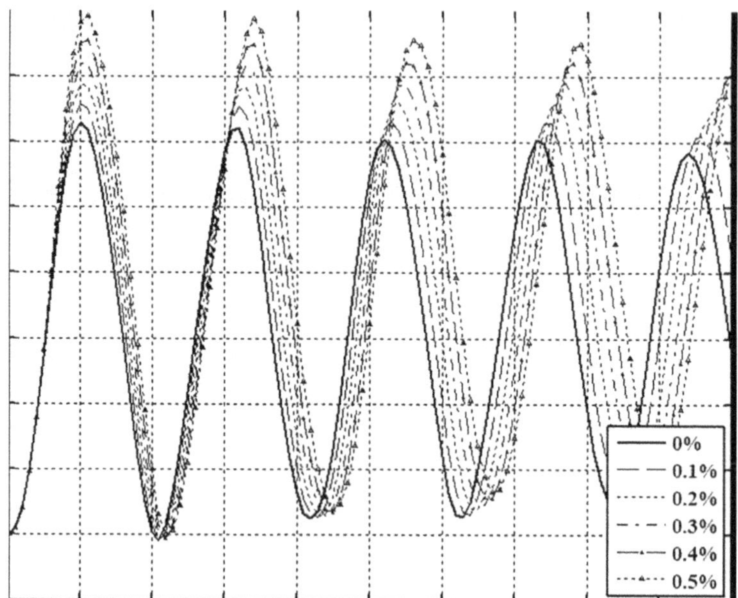

FIGURE 5.11 (Continued)

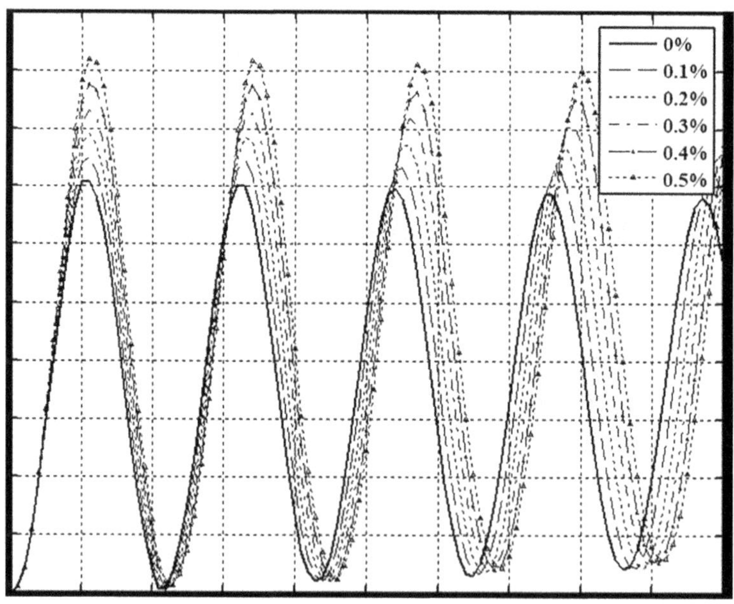

FIGURE 5.12 Nonlinear dynamic response of a simply supported anti-symmetric cross-ply $(0^o / 90^o / 0^o / 90^o)$ square substrate plate undergoing active constrained layer damping (for $a / h = 300$, $Q = 100$) using AFC for various moisture conditions: a) under passive mode ($K_d = 0$); b) under active mode ($K_d = 100$).

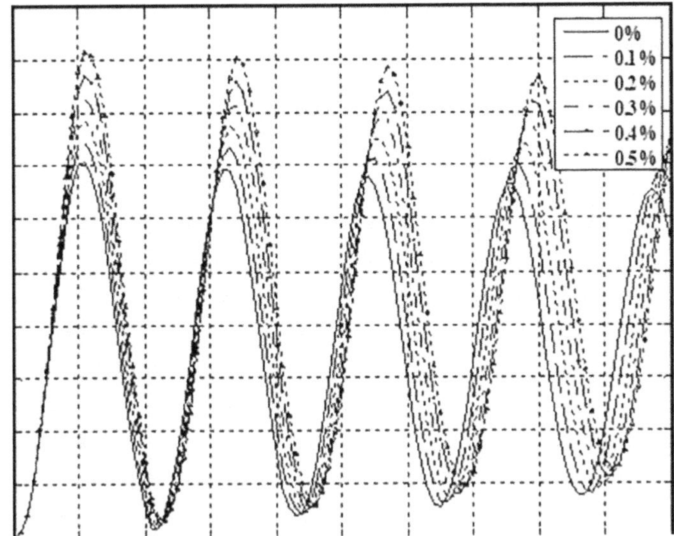

FIGURE 5.12 (Continued)

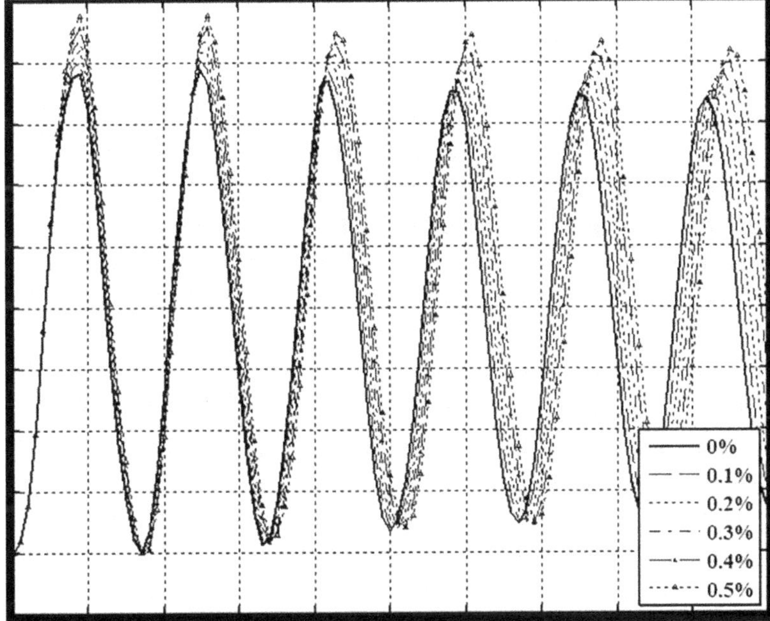

FIGURE 5.13 Nonlinear dynamic response of a simply supported anti-symmetric angle-ply $(-45^o / 45^o / -45^o / 45^o)$ square substrate plate undergoing active constrained layer damping (for $a / h = 300$, $Q = 100$) using AFC for various moisture conditions: a) under passive mode ($K_d = 0$); b) under active mode ($K_d = 100$).

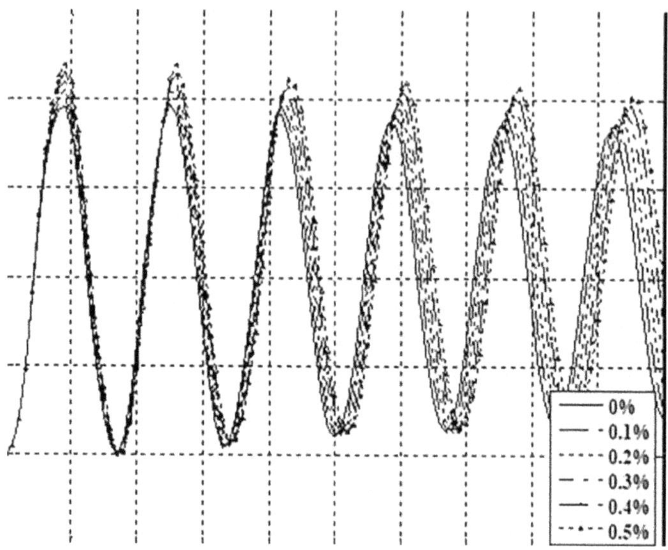

FIGURE 5.13   (Continued)

5.5   CONCLUSIONS

The performance of laminated composite plates using the ACLD treatment using PFRC/AFC material under the influence of moisture is examined. Numerical results revealed that the amplitude of the vibration significantly increases with an increase in moisture content in the environment, which is evident from the results. The experiment is conducted for various percentages of moisture content in the composite plate with a considerable increase in the amplitude of the vibration with the increase in moisture. The graphs reveal that the experiments done on symmetric cross-ply (0°/90°/0°), (0°/90°/0°/90°) and (−45°/45°/−45°) for the PFRC and AFC square substrates are compared, and it is observed that the deflection in the PFRC substrate is more than that of the AFC substrate, with a considerable increase in the moisture content. Hence the performance of the AFC substrate is considerably better than that of the AFC material in reducing vibrations at different moisture percentages.

REFERENCES

1. M K Rath and S K Sahu, Vibration of woven fiber laminated composite plates in hygrothermal environment, *Journal of Vibration and Control* 18(13), 2012, pp. 1957–1970.
2. B P Patel, M Ganapathi, and D P Makhecha, Hygrothermal effects on the structural behavior of thick composite laminates using higher-order theory, *Composite Structures* 56, 2002, pp. 25–34.

3. S K Singh, and A Chakrabarti, Hygrothermal analysis of laminated composite plates by using efficient higher order shear deformation theory, *Journal of Solid Mechanics* 3(1), 2011, pp. 85–95.

4. X Li, K Yu, J Han, H Song, and R Zhao, Buckling and vibro-acoustic response of the clamped composite laminated plate in thermal environment, *International Journal of Mechanical Sciences* 119, 2016, pp. 370–382.

5. L S Ramachandra and S K Panda, Dynamic instability of composite plates subjected to non-uniform in plane loads, *Journal of Sound and Vibration* 331(1), 2012, pp. 53–65.

6. H R Ovesy and J Fazilati, Parametric instability analysis of moderately thick FGM cylindrical panels using FSM, *Computers and Structures* 108–109, 2012, pp. 135–143.

7. Y Tang and X Wang, Buckling of symmetrically laminated rectangular plates under parabolic edge compressions, *International Journal of Mechanical Sciences* 53(2), 2011, pp. 91–97.

8. M Rafiee, X Q He, and K M Liew, Non-linear dynamic stability of piezoelectric functionally graded carbon nanotube-reinforced composite plates with initial geometric imperfection, *International Journal of Non-Linear Mechanics* 59, 2014, pp. 37–51.

9. A Garg and H D Chalak, A review on analysis of laminated and sandwich structures under hygrothermal conditions, *Thin-Walled Structures* 142, 2019, pp. 205–226.

10. S Yung Lee, J L Jang, J S Lin, and C J Chou, Hygrothermal effects on the linear and nonlinear analyses of composite plates, *Composite Structures* 21(1), 1991, pp. 177–186.

11. J Shivakumar, M H Ashok, and V Khadakbhavi, Performance analysis of geometrically nonlinear static deformations of laminated composite plates using PFRC/AFC materials, *Materials Today: Proceedings* 5(13), 2018, 27254–27259.

12. J Shivakumar, M H Ashok, V Khadakbhavi, S Pujari, and S Nandurkar, Performance analysis of smart laminated composite plate integrated with distributed AFC material undergoing geometrically nonlinear transient vibrations, *IConAMMA-2017 IOP Publishing IOP Conference Series: Materials Science and Engineering* 310, 2018, 012100. doi:10.1088/1757-899X/310/1/012100

13. M H Ashok, J Shivakumar, S Nandurkar, V Khadakbhavi, and S Pujari, Geometrically nonlinear transient vibrations of actively damped anti-symmetric angle ply laminated composite shallow shell using active fiber composite (AFC) actuators, *IConAMMA-2017 IOP Publishing IOP Conference Series: Materials Science and Engineering* 310, 2018, 012101. doi:10.1088/1757-899X/310/1/012101

14. V Khadakbhavi, M H Ashok, J Shivakumar, S Pujari, and S Nandurkar, Comparative study on performance of controlling geometric nonlinearity of laminated smart composite plate using piezoelectric fiber reinforced composite & active fiber composite material, *IConAMMA2018 IOP Conference Series: Materials Science and Engineering* 577, 2019, 012164. doi: 10.1088/1757-899X/577/1/012164

15. S Nandurkar, M H Ashok, J Shivakumar, V Khadakbhavi, and S Pujari, Comparative investigation on nonlinear transient vibrations of laminated composite shallow shell with active constrained layer damping treatment using PFRC and AFC patch material, *IConAMMA 2018 IOP Conference Series: Materials Science and Engineering* 577, 2019, 012147. doi: 10.1088/1757-899X/577/1/012147

16. K S S Ram, and P K Sinha, Hygrothermal effects on the bending characteristics of laminated composite plates, *Computers & Structures* 40(4), 1991, pp. 1009–1015.

17. S Joshan Yadwinder, S Santapuri, and N Grover, Analysis of laminated piezoelectric composite plates using an inverse hyperbolic coupled plate theory, *Applied Mathematical Modelling* 82, 2020, pp. 359–378.

18. A Chanda and R Sahoo, Analytical modeling of laminated composite plates integrated with piezoelectric layer using Trigonometric Zigzag theory, *Journal of Composite Materials* 54(29), 2020.

19. S Ghosh, S Agrawal, A K Pradhan, and M K Pandit, Performance of vertically reinforced 1–3 piezo-composites for active damping of smart sandwich beams, *Journal of Sandwich Structures & Materials* 17(3), 2015.

20. Y S Joshan, S Santapuri, and A Srinivasa, Finite element modeling and analysis of low symmetry piezoelectric shells for design of shear sensors, *International Journal of Mechanical Sciences* 210, 2021, p. 106726.

21. Y M Ghugal, and S K Kulkarni, Flexural analysis of cross-ply laminated plates subjected to nonlinear thermal and mechanical loadings, *Acta Mechanica* 224(3), 2013, pp. 675–690.

22. Y M Ghugala and S K Kulkarni, Thermal response of symmetric cross-ply laminated plates subjected to linear and non-linear thermo-mechanical loads, *Journal of Thermal Stresses* 36(5), 2013, pp. 466–479.

23. B K Shravankumar and A Ghosh, Geometrically non-linear bending analysis of piezoelectric fiber reinforced composite (MFC/AFC) cross-ply plate under hygro-thermal environment, *Journal of Thermal Stresses* 36(12), 2013, pp. 1255–1282.

24. N ur Rahmana and M Naushad Alama, Structural control of piezoelectric laminated beams under thermal load, *Journal of Thermal Stresses* 38(1), 2015, pp. 69–95.

25. X Zhao, F J N Iegaink, W D Zhu, and Y H Li, Coupled thermo-electro-elastic forced vibrations of piezoelectric laminated beams by means of Green's functions, *International Journal of Mechanical Sciences* 156, 2019, pp. 355–369.

26. L Zhou, B Nie, S Ren, K K Żur, and J Kim, On the hygro-thermo-electro-mechanical coupling effect on static and dynamic responses of piezoelectric beams, *Composite Structures* 259, 2021, 113248.

27. T Liu, C Li, C Wang, W Hu, and T Q Bui, Geometrically nonlinear isogeometric analysis of smart piezoelectric FG plates considering thermal effects of piezoelectric stress and dielectric constants, *Composite Structures* 266, 2021, 113795.

28. M C Ray, T Shivakumar, Active constrained layer damping of geometrically nonlinear transient vibrations of composite plates using piezoelectric fiber reinforced composite, *Thin-Walled Structures Journal* 47, 2009, pp. 178–189.

29. S K Sarangi and M C Ray, Smart control of nonlinear vibrations of laminated plates using active fiber composites, *International Journal of Structural Stability and Dynamics* 12(6), 2012, doi:10.1142/S0219455412500502

6 Investigating the Impact of Parameters on Electrical Discharge Treatment of ZE41A Magnesium Alloy Utilizing a WC/Ni Powder Composite Electrode

P. Senthil Kumar, U. Elaiyarasan,
V. Satheeshkumar, and C. Senthil Kumar

6.1 INTRODUCTION

Surface-modified materials are becoming more and more essential in technical applications like automobiles, airplanes, and aerospace. Many surface modification techniques have been applied on workpiece materials to improve surface properties [1]. Surface modification operations can be divided into two categories: material modification through deposition or coating and alloy creation through casting and powder metallurgy. Today, changed surfaces with coatings or depositions are extensively used in a variety of application domains [2]. To improve the surface qualities, a variety of surface modification techniques are used, including HVOF, CVD, PVD, and thermal spraying. Due to their accuracy, these methods are widely applied across numerous industries. To improve the properties, however, specialized equipment is required to manage the coating's features [3]. The workpiece's surface qualities can be improved using the electro discharge deposition (EDD) technique, which can help solve this issue. The widely recognized non-traditional machining technique known as EDM is used to precisely and accurately machine hard materials. In addition to being used for machining [4], EDM is also used to evenly apply a coating layer of the required wear-resistant

DOI: 10.1201/9781003397465-6

material to the surface. In this method, the electrode particles are coated on the workpiece through a continuous electric spark between the powder compact electrode and the base materials. The electrode materials are melted during this process and applied as a coating layer to the surface [5]. Numerous studies have been conducted on various workpiece and electrode materials; nevertheless, there has been very little optimization of this type of modification technique.

Suzuki et al. [6] investigated the surface modification of high-speed steel with adhesion TiC particles using EDC. The adoption of powder metallurgy processed (PM) electrodes in EDM has emerged as a viable alternative tooling choice, offering enhanced surface characteristics. Ahmed [7] modified the surface of an aluminium alloy with the aid of a powder metallurgy electrode using EDL. Different EDM factors have been employed to control the coating performance. EDC processed with a higher current rate and pulse on time increased deposition rate (DR) and decreased SR. Similarly, Gill et al. [8] investigated the influence of conventional copper electrodes and powder compact electrodes (Cu-Cr-Ni) on the DR and SR. As pulse current and pulse on time increases, the DR and layer thickness (LT) increase. The surface topography of the coating was characterized using scanning electron microscopy (SEM). The maximum DR and LT were obtained at the powder composite electrode. This could be due to controlling the density of the electrode.

From the literature, the various aspects of EDM were investigated with a focus on the use of powder metallurgy processes. The deposition of electrode particles in single pulse discharges was analysed. PM green compacts for surface modification of aluminium work pieces were employed. The effects of semi-sintered electrodes properties having the advantages of less SR in duplex stainless coatings. The microhardness of a composite tool coated using EDM was explored, with an emphasis on material deposition from the electrode and dielectric onto the machined surface. These studies collectively contribute to understanding and optimizing EDM processes for enhanced surface properties. In this work, a magnesium alloy was modified with a Tungsten Carbide/ Nickel (WC/Ni) coating using the EDM process.

6.2 MATERIALS AND METHODS

Magnesium alloys are lightweight structural materials that are widely used in a variety of industries, including aerospace and automotive. The base material used in this work is a ZE41A magnesium alloy, whose chemical makeup is presented in Table 6.1.

WC and Ni are the chosen as coating materials for the electrodes. The chosen electrode powders are mixed with WC70 Ni30 (wt.%) for three hours. After that, the combined powders are compressed for ten minutes with three distinct loads in a hydraulic press utilizing a punch and die system. The die and upper and lower punch were fabricated with diameter of 10 mm. After compaction was complete, the solid object inside the die was smoothly removed with the aid of paraffin oil.

TABLE 6.1

Chemical Elements of ZE41A Mg Alloy

Element	Mg	Zn	Ce	Zr	Mn	Cu	Ni
wt.%	94	3.5–5	0.5–1.75	0.40–1.0	0.15	0.10	0.010

FIGURE 6.1   WC/Ni compacted electrode.

Finally, the cylindrical composite electrode was ready for the coating process and is depicted in Figure 6.1.

In this work, EDC experiments employ ordinary EDM (Figure 6.2; model: 5530 EDM E-series). As a dielectric fluid, EDM oil was used, which controls the temperature generated between the electrode and workpiece. The central composite design (CCD) in RSM was adopted as a design tool in this work. EDM factors and ranges were fixed to conduct the investigation and are displayed in Table 6.2. The workpiece was machined into a square form (20 × 20 mm) using a grinder machine and polished to a high gloss using an emery sheet abrasive with 1500 grit before the trials were conducted. The hardness of the workpiece material was about 68 HB. Every experiment had a set duration of ten minutes and was carried out in accordance with the design created using the RSM tool. With an accuracy of 0.001 g for the weighing machine, the weight of the base metal and electrode was measured using a weighing instrument (model: SF400D).

6.3   DESIGN OF EXPERIMENTS

RSM is the term used to describe a group of mathematical methods used for problem analysis and modelling where mathematical models have been used to establish responses [9]. The goal is to minimize the number of tests and pinpoint the factor that influences the coating process. EDM factors were assessed using the empirical model mathematical model. Equation 6.1 gives a second-order polynomial equation.

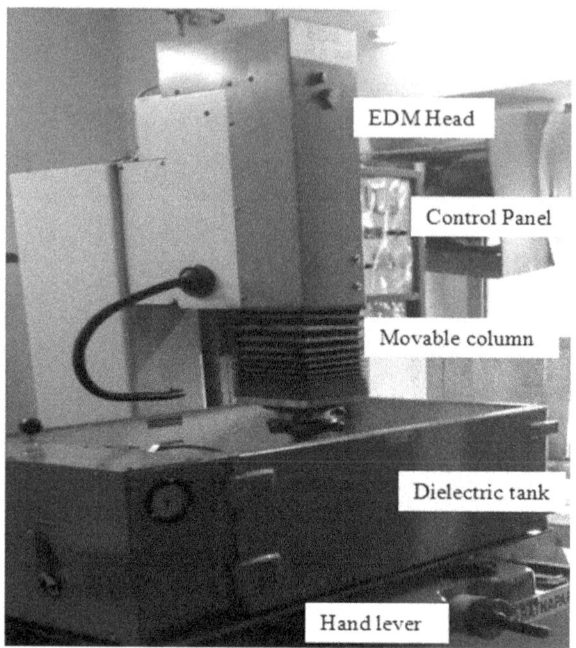

FIGURE 6.2 EDM machine (model: 5530 EDM E-series).

TABLE 6.2

EDM Factors and Ranges

Factors	Ranges		
	Lower	Intermediate	Higher
Compaction load [CL]	150	175	200
Current [I]	2	3	4
Pulse on time [T_{on}]	50	70	90

$$Z_u = a_o + \sum_{i=1}^{k} a_i x_i + \sum_{i=1}^{k} a_{ii} x^2{}_i + \sum_{j>1}^{k} a_{ij} x_i x_j \tag{6.1}$$

where Zu = responses, a_i = linear term of x_i, a_{ii} = square formula of x_j, and b_{ij} = interaction formula of x_i and x_j.

In the previous formulated equation, first, second, third, and fourth terms are a constant, linear, square, and interaction, respectively. The tests were conducted in accordance with the design, and Table 6.3 displays the results. Minitab 17.0 was used to do the statistical analysis.

TABLE 6.3
Experimental Values

	Factors			Responses	
S. No	Compaction Load (MPa)	Current (A)	Pulse on Time (µs)	DR (g/min)	EWR (g/min)
1	150	2	50	0.0302	0.1617
2	200	2	50	0.0692	0.0567
3	150	4	50	0.1082	0.2227
4	200	4	50	0.0722	0.1577
5	150	2	90	0.1462	0.2997
6	200	2	90	0.0602	0.2247
7	150	4	90	0.2959	0.5907
8	200	4	90	0.1181	0.5603
9	150	3	70	0.1287	0.4677
10	200	3	70	0.0817	0.3677
11	175	2	70	0.0799	0.1798
12	175	4	70	0.1471	0.3758
13	175	3	50	0.1077	0.1204
14	175	3	90	0.1878	0.3898
15	175	3	70	0.1443	0.3318
16	175	3	70	0.1376	0.3413
17	175	3	70	0.1329	0.3301
18	175	3	70	0.1386	0.3344
19	175	3	70	0.1376	0.3268
20	175	3	70	0.1318	0.3315

6.3.1 MATHEMATICAL MODEL

Mathematical expressions are created for DR and EWR to study the effect of parameters and are given in Equations 6.2 and 6.3.

$$\begin{aligned} DR = {}& 2.13061 + 0.01899*k + 0.27849*l + 0.004510*m \\ & - 0.00004\left(k*k\right) - 0.0193\left(l*l\right) + 0.00015\left(m*m\right) \\ & + 0.0073\left(k*l\right) - 0.0007\left(k*m\right) + 0.0089\left(l*m\right) \end{aligned} \tag{6.2}$$

$$R^2 = 96.06$$

$$\begin{aligned} EWR = {}& 3.57896 - 0.04933*k + 0.15750*l + 0.02327*m \\ & + 0.00015\left(k*k\right) - 0.05563\left(l*l\right) - 0.00017\left(m*m\right) \\ & + 0.00053\left(k*l\right) + 0.00003\left(k*m\right) + 0.00320\left(l*m\right) \end{aligned} \tag{6.3}$$

$$R^2 = 97.10$$

where k = compaction load, l = discharge current, and m = pulse on time.

TABLE 6.4
ANOVA for DR

Source	DF	Seq SS	Adj SS	Adj MS	F	P
Regression	9	0.05905	0.05842	0.00672	135.74	0.000
Linear	3	0.04074	0.01040	0.00364	70.27	0.000
k	1	0.01036	0.00353	0.00353	67.98	0.000
l	1	0.01356	0.00265	0.00265	49.36	0.000
m	1	0.01859	0.00067	0.00067	7.24	0.013
Square	3	0.00483	0.00419	0.00157	26.33	0.000
k*k	1	0.00365	0.00192	0.00192	33.67	0.000
l*l	1	0.00118	0.00098	0.00098	13.84	0.002
m*m	1	0.00178	0.00115	0.00115	17.41	0.001
Interaction	3	0.01527	0.01464	0.00505	100.29	0.000
k*l	1	0.00437	0.00374	0.00374	72.34	0.000
k*m	1	0.00979	0.00916	0.00916	187.47	0.000
l*m	1	0.00290	0.00226	0.00226	41.06	0.000
Residual error	10	0.00136	0.00073	0.00031		
Lack-of-fit	5	0.00126	0.00063	0.00033	2.14	0.08
Pure error	5	0.00099	0.00036	0.00028		
Total	19	0.05952				

The F-test and ANOVA test were performed to verify the goodness of fit. The lack of fit suggests the mathematical model holds statistical significance for both DR and EWR. As seen by the R^2 values of 96.06 for EWR and 97.10 for DR, the regression model offers substantial support for the features. In the ANOVA table, the developed model is considered statistically significant if the P value is less than 0.05. Tables 6.4 and 6.5 list the ANOVA table for DR and EWR. At a 95% confidence level, the typical percentage of the F distribution is 4.06, and for a 99% confidence level, it is 7.87. The F points of lack of fit for DR and EWR are 2.14 and 3.25, respectively. These values are lower than the typical F values. Thus, the models are adequate.

Figures 6.3 and 6.4 display the normal probability graphs for DR and EWR, respectively, aiming to assess the distribution of residuals. The plots demonstrate that the residuals align along a straight line, indicating the regression model's adequacy and normal distribution of errors.

6.4   RESULTS AND DISCUSSION

The main effect plots for DR and EWR are presented in Figures 6.5 and 6.6. It is shown that DR and EWR exhibit higher values at 150 MPa, 2 A, and 50 µs. The explanation for this trend is attributed to the limited mechanical interaction among powder particles at low compaction, leading to a substantial

TABLE 6.5
ANOVA for EWR

Source	DF	Seq SS	Adj SS	Adj MS	F	P
Regression	9	0.36613	0.362802	0.04583	783.05	0.000
Linear	3	0.29809	0.028098	0.01432	168.43	0.000
k	1	0.01996	0.024788	0.02805	436.21	0.000
l	1	0.10282	0.003656	0.00692	24.01	0.003
m	1	0.18705	0.010564	0.01382	158.76	0.000
Square	3	0.04551	0.042183	0.01901	260.01	0.000
k*k	1	0.00599	0.022269	0.02553	387.07	0.000
l*l	1	0.02869	0.010928	0.01419	165.86	0.000
m*m	1	0.02257	0.019244	0.02250	328.08	0.000
Interaction	3	0.03427	0.030942	0.01527	186.92	0.000
k*l	1	0.00677	0.003435	0.00670	19.71	0.003
k*m	1	0.00639	0.003062	0.00632	12.43	0.009
l*m	1	0.03286	0.029525	0.03279	528.61	0.000
Residual error	10	0.00638	0.003053	0.00585		
Lack-of-fit	5	0.00626	0.002932	0.00588	3.25	0.111
Pure error	5	0.00599	0.002661	0.00582		
Total	19	0.36664				

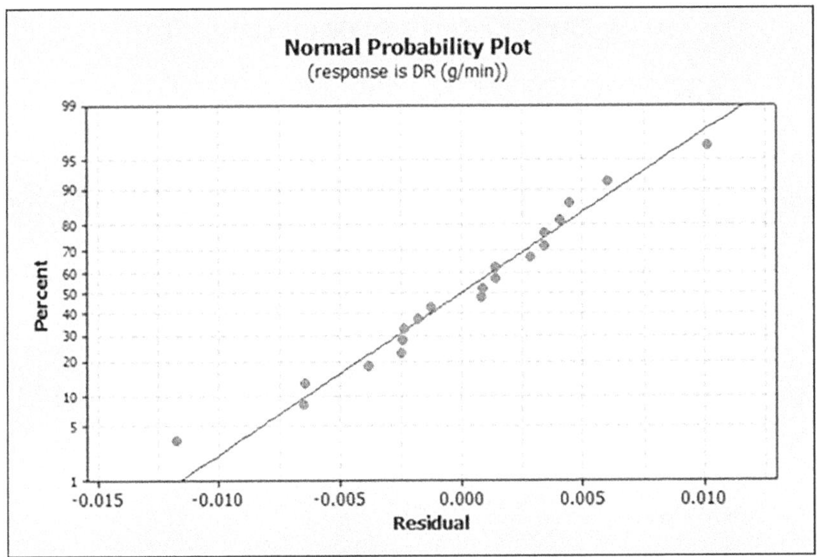

FIGURE 6.3 Normal probability DR.

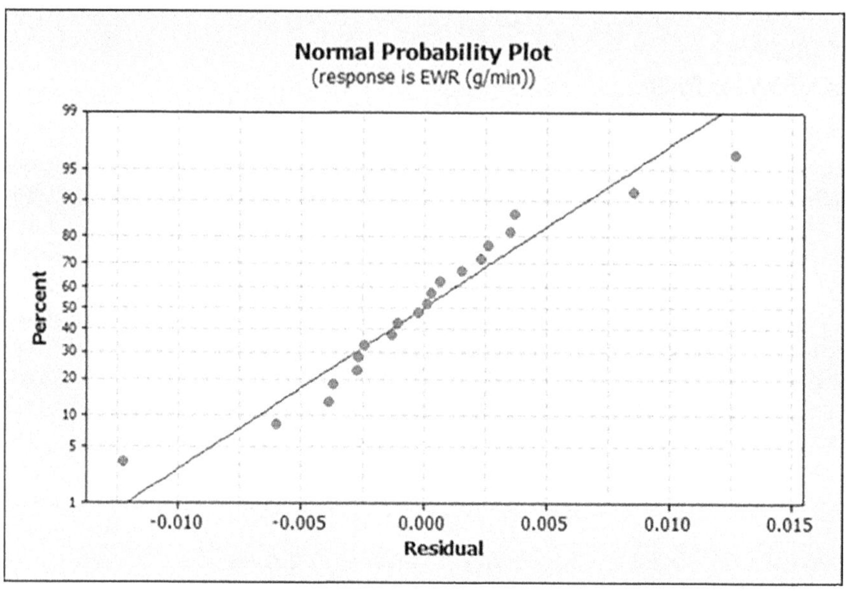

FIGURE 6.4   Normal probability for EWR.

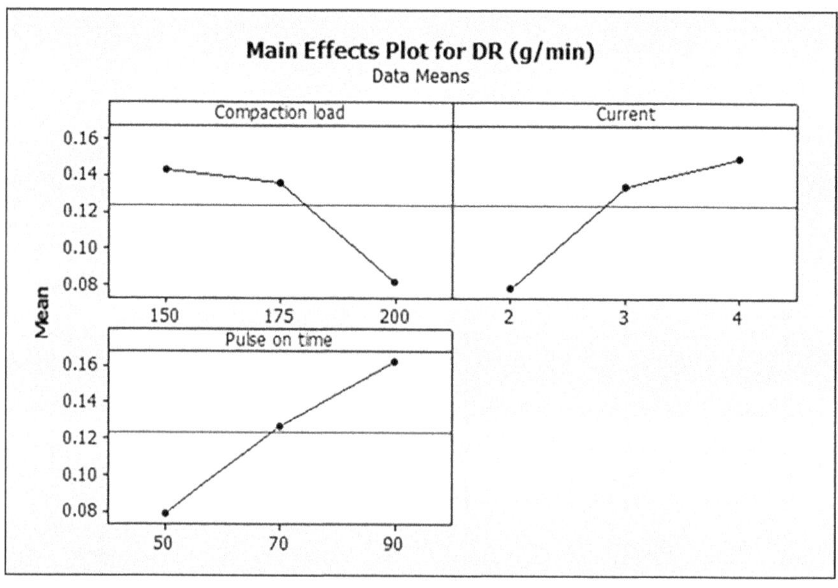

FIGURE 6.5   Main effect plot for DR.

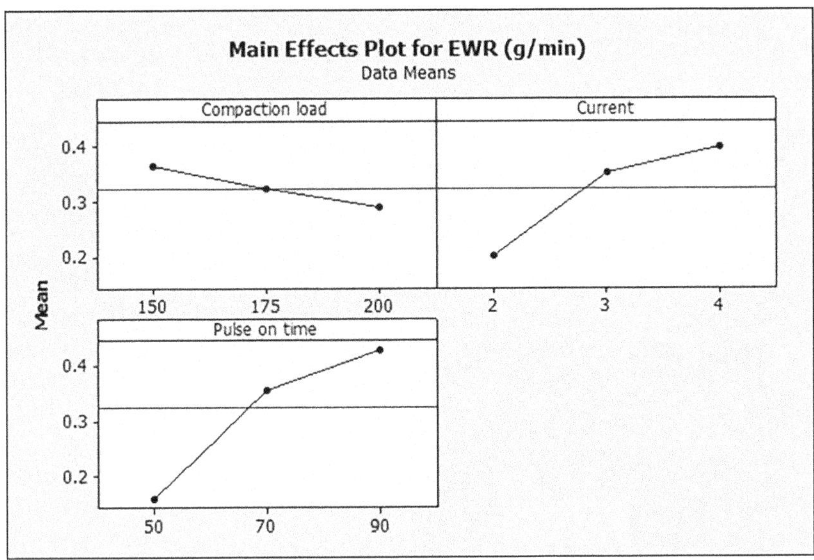

FIGURE 6.6 Main effect plot for EWR.

disintegration of particles from the composite electrode during the EDD process [10]. Consequently, a notable amount of powder particles are deposited on the workpiece surface. In a similar vein, Walia showed that the surface deposited using an electrode created at low compaction stress has a high DR. Conversely, electrodes prepared with large compaction loads have extremely high levels of particle bonding. Thus, a small amount of the electrode material was applied to the workpiece [11]. The SEM microstructure of the deposited surface was examined under various circumstances, as illustrated in Figure 6.7a–c. In Figure 6.7a, the microstructure displays the pores on the surface formed under the parameters of 150 MPa, 4 A, and 90 µs. Similarly, Figure 6.7c depicts globules observed under the conditions of 200 MPa, 2 A, and 50 µs. Additionally, Figure 6.7b shows craters on the surface formed at 175 MPa, 3 A, and 70 µs.

The coating thickness in the surface ranges from 78 to 119 µm. Specifically, at the conditions of 150 MPa, 4 A, and 90 µs, the deposited surface exhibits a high surface roughness of 6.89 µm, primarily due to the formation of numerous pores under these specific parameters. At 200 MPa, 2 A, and 50 µs, the minimum surface roughness of 4.35µm was attained due to a homogenous distribution of tiny globules. Figure 6.8 displays an electrode material that is surface presented and whose validity is verified by an energy dispersive spectroscopy (EDS) peak map. Additionally, there is an observed trend that as the pulse current increases, the spark density also increases, leading to higher values for both DR and EWR.

Poor electrode material erosion results from insufficient electric spark production at lower current settings. Indeed, the heightened spark intensity resulting from increased current contributes to the elevation of both DR and EWR. Similarly, the

FIGURE 6.7   SEM: a) 150 MPa, 4 A, 90 µs; b) 175 MPa, 3 A, 70 µs; and c) 200 MPa, 2 A, 50 µs.

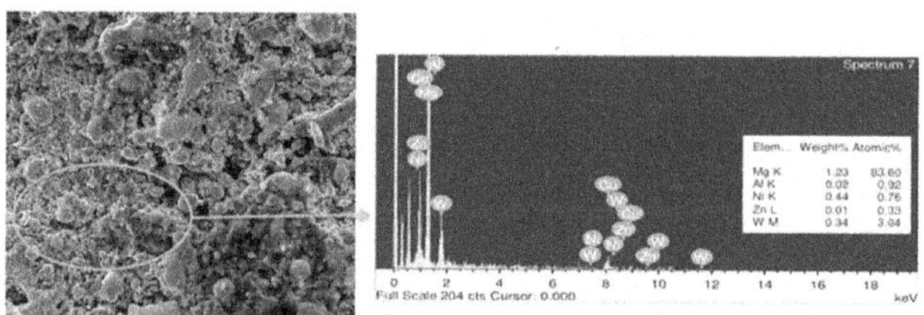

FIGURE 6.8   EDS plot at 150 MPa, 4 A, and 90 µs.

investigation indicates that EWR experiences an increase in tandem with the rise in current. Due to the formation of larger and deeper craters, a substantial number of particles undergo degradation and get deposited over the surface, leading to an increase in the roughness of the resulting surface [12]. Both DR and EWR exhibit an upward trend with an increase in pulse on time. The investigation affirms that higher pulse on time, coupled with a larger plasma channel diameter, results in the disintegration of a significant quantity of electrode particles. Consequently, these materials are coated on the workpiece through the generated plasma channel. Insufficient discharge energy causes the electrode materials to melt and evaporate, resulting in very little electrode material migration at low pulses [13]. Consequently, very little electrode material comes into contact with the surface. DR and EWR fall as a result. More uniform deposition with better surface roughness resulted under these conditions.

6.4.1 MICROHARDNESS

Microhardness, a crucial characteristic of the deposited surface, was assessed using a Vickers hardness tester. Each sample underwent three measurements, and the average of these values was used to determine the microhardness [14]. Figure 6.9 displays the microstructure of the coating, highlighting the microhardness values obtained in various areas. The deposited surface has an average microhardness value of 772 HV. As mentioned earlier, the microhardness values are higher at low compaction due to the bulk and lump deposition processes. As a result, the hard materials that were deposited resist indentation during hardness tests. Hence, MH is high with low compaction loads. On the other hand, because of the huge indentation diameter, the MH value is minimal under high compaction loads. As a result, the MH falls as the compaction load rises.

6.5 OPTIMIZATION

At present, the manufacturing industry has used a number of optimization techniques to improve machining parameters. However, some strategies make it

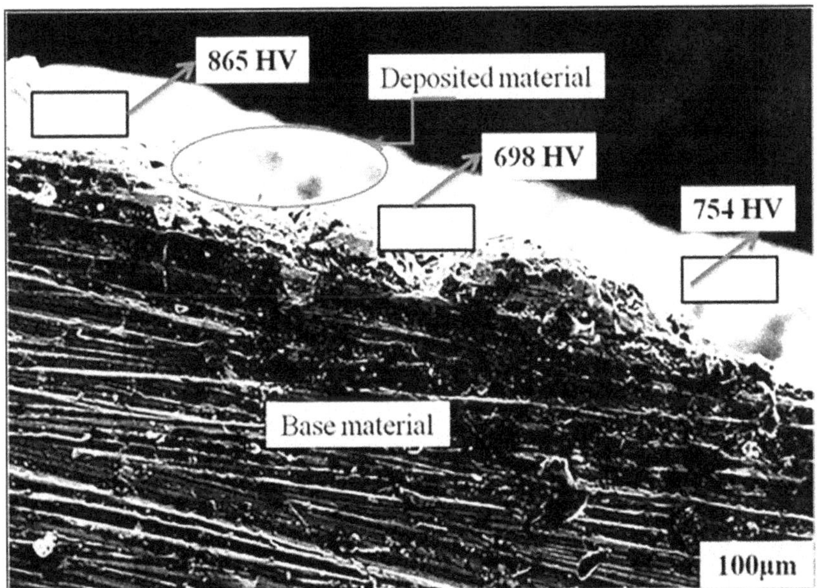

FIGURE 6.9 Microstructure of the microhardness.

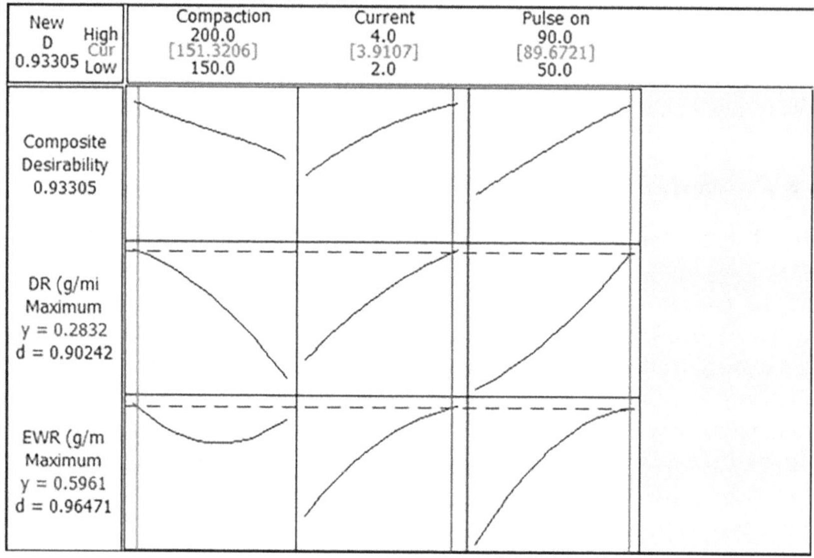

FIGURE 6.10   Optimization plot.

impossible to alter the parameters while getting numerous responses at once [15]. Response surface methodology, on the other hand, is a significant tool for multi-responder parameter optimization. Creating a desirability function di from the response is the main optimization technique. By increasing the composite desirability, the optimal value for each parameter can be identified. The outcome of the study is to increase the desirability of the responses.

$$0 < d_i < 1 \tag{6.4}$$

In this work, maximizing the value of DR and EWR is the main objective of the optimization. The optimization process involves setting target values, and parameters are adjusted to achieve the highest and lowest experimental values. In Figure 6.10, it can be observed that the maximum DR of 0.283 g/min and EWR of 0.5961 were attained at 151 MPa, 4 A, and 89 µs. The outcomes of the validation tests for both DR and EWR are shown in Figure 6.10. The desirability values of DR and EWR were 0.929 and 0.938, respectively, and the composite optimization desirability was 0.933, which is extremely close to 1. Every parameter is therefore within the operational range.

6.7   CONCLUSION

The current study focused on the parametric optimization of EDD for a magnesium alloy using RSM. The objective was to deposit a WC/Ni coating on the surface of the magnesium alloy through a controlled process involving various factors.

Under intermediate load conditions (175 MPa), there is an observed increase in both DR and EWR with a simultaneous increase in pulse current and on time. This is because the coating zone's local temperature rises as a result of the higher spark density. As a result, a significant amount of material is released from the electrode and settled onto the workpiece. On the other hand, it falls as the compaction load rises. Higher bonding strength can be the cause of this. The electrode solidified into a fully bonded state when it was prepared with a high compaction load. The pace at which the electrode erodes is lower even with high pulse energy delivered. Consequently, under heavy compaction stress, DR and EWR dropped. The three most crucial study factors are the compaction load, current, and pulse on time. An empirical model was developed for DR and EWR in order to evaluate the parametric effect. The DR and EWR responses were optimized by RSM. The optimal parameters were a current of 4 A, pulse on time of 90 µs, and compaction load of 150 MPa.

REFERENCES

1. Barshilia, H.C. 2021. Surface modification technologies for aerospace and engineering applications: current trends, challenges and future prospects. *Transactions of the Indian National Academy of Engineering*, 6(2), pp. 173–188.
2. Liu, X., Paul, K.C. and Ding, C. 2004. Surface modification of titanium, titanium alloys, and related materials for biomedical applications. *Materials Science and Engineering: Reports*, 47(3–4), pp. 49–121.
3. Mann, B.S., Arya, V., Maiti, A.K., Rao, M.U.B. and Joshi, P. 2006. Corrosion and erosion performance of HVOF/TiAlN PVD coatings and candidate materials for high pressure gate valve application. *Wear*, 260(1–2), pp. 75–82.
4. Volosova, M.A., Okunkova, A.A., Fedorov, S.V., Hamdy, K. and Mikhailova, M.A. 2020. Electrical discharge machining non-conductive ceramics: combination of materials. *Technologies*, 8(2), p. 32.
5. Zhao, H., Gao, C., Wu, X.Y., Xu, B., Lu, Y.J. and Zhu, L.K. 2019. A novel method to fabricate composite coatings via ultrasonic-assisted electro-spark powder deposition. *Ceramics International*, 45(17), pp. 22528–22537.
6. Rajamurugan, T.V., Rajaganapathy, C., Jani, S.P., Gurram, C.S., Allasi, H.L. and Damtew, S.Z. (2022). Analysis of drilling of coir fiber-reinforced polyester composites using multifaceted drill bit. *Advances in Materials Science and Engineering*, 2022(1), 9481566.
7. Ahmed, A. 2016. Deposition and analysis of composite coating on aluminum using Ti–B4C powder metallurgy tools in EDM. *Materials and Manufacturing Processes*, 31(4), pp. 467–474.
8. Gill, A.S. and Kumar, S. 2014. Surface roughness evaluation for EDM of En31 with Cu-Cr-Ni powder metallurgy tool. *International Journal of Industrial and Manufacturing Engineering*, 8(7), pp. 1308–1313.
9. Mia, M. 2018. Mathematical modeling and optimization of MQL assisted end milling characteristics based on RSM and Taguchi method. Measurement, 121, pp. 249–260.

10. Elaiyarasan, U., Satheeshkumar, V. and Senthilkumar, C. 2018. Experimental analysis of electrical discharge coating characteristics of magnesium alloy using response surface methodology. *Materials Research Express*, 5(8), p. 086501.
11. Elaiyarasan, U., Satheeshkumar, V. and Senthilkumar, C. 2021. Effect of parameters on microstructure of electrical discharge coated ZE41A magnesium alloy with tungsten carbide-copper composite electrode. *Surface Topography: Metrology and Properties*, 9(2), p. 025006.
12. Krella, A. and Marchewicz, A. 2023. Effect of mechanical properties of CrN/CrCN coatings and uncoated 1.402 stainless steel on the evolution of degradation and surface roughness in cavitation erosion. *Tribology International*, 177, p. 107991.
13. Rajkumar, T., Radhakrishnan, K., Rajaganapathy, C., Jani, S.P. and Ummal Salmaan, N. 2022. Experimental investigation of AA6063 welded joints using FSW. *Advances in Materials Science and Engineering*, 2022(1), 4174210.
14. Karatas, O., Gul, P., Akgul, N., Celik, N., Gundogdu, M., Duymus, Z.Y. and Seven, N. 2021. Effect of staining and bleaching on the microhardness, surface roughness and color of different composite resins. *Dental and Medical Problems*, 58(3), pp. 369–376.
15. Köksal, G., Batmaz, I. and Testik, M.C. 2011. A review of data mining applications for quality improvement in manufacturing industry. *Expert Systems with Applications*, 38(10), pp. 13448–13467.

7 Experimental Investigations of Drilling Characteristics of Derlin Using Hard Carbide and HSS Drill Bits— Taguchi and Decision Tree Algorithm

G Jeeva Kumar, R Arun, Masi Periyasaame R, Balasundaram R, Jafrey Daniel James D, Karthik Pandiyan G, and Muthu Chozha Rajan B

7.1 INTRODUCTION

Composite materials offer benefits such as improved strength-to-weight ratio, improved thermal properties, lower specific gravity, and high damping resistance. Polymer products are employed in a wide range of industries, including automotive and aerospace [1]. Polymers are replacing metals in a variety of industrial equipment due to their improved corrosion resistance. While these benefits make composites a viable alternative to metals, there are a number of issues with regard to machining [2].

Despite the fact that numerous manufacturing techniques can manufacture composites in a near-net form, composite machining is inevitable. Drilling, milling, orthogonal cutting, and finishing operations may indeed be used to machine composites. A wedge-shaped cutting tool is used in the oblique cutting process to cut materials in the form of discontinuous chips. Turning is a method in which the work piece rotates and the tool travels around the axis of the work piece [3].

Drilling is a process that involves using different types of drill bits to create holes in a part. Drilling is one of the most widely used machining processes for polymer materials, and it is used to assemble the parts into usable items [4]. The accuracy of the diameter of the hole, surface finish of the drilled hole, material delamination, and circularity can all be used to quantify the

DOI: 10.1201/9781003397465-7

efficiency of the drilled hole. Drilling input parameters must be carefully chosen to ensure the accuracy of the drilled hole. Surface roughness is the most serious issue among the aforementioned issues because it determines the machined surface's quality. The choice of matrix and fibre is crucial for effective polymer machining [5].

Several researchers have previously experimented with drilling polymer materials by changing the tool material, changing the cutting tool, and choosing various tool angles. For machining of glass fibre-reinforced composites with a tool made of polycrystalline diamond, Palanikumar et al. [6] used response surface and Taguchi methodology to optimize surface roughness. They came to the conclusion that feed rate and cutting pace are the most important variables. Prakash et al. [7] investigated the impact of spindle speed, drill diameter, and feed rate on the drilling of medium-density fibre board using a Taguchi L27 orthogonal array, grey relational analysis, and analysis of variance (ANOVA). Diaz Alavarez et al. [8] investigated the drilling properties of polylactic acid reinforced with cotton, flax, and jute produced by compression moulding. The delamination factor was used to investigate the output characteristics in relation to drilling speed, twist drill angle, and feed rate. When the cutting speed and feed rate were increased, the delamination factor was found to be lower. The fibre was cut without fraying when the cutting speed was increased. In the case of delamination, the most important factor was twist drill angle, while the least important factor was drill diameter.

Pandey et al. [9] used grey relational analysis and a fuzzy-based algorithm to optimize bone drilling and found that the feed rate was the most important factor. When drilling MMC, Gowda et al. [10] looked at the impacts of cutting parameters on surface roughness, cylindricity, and circularity. Kaviarasan et al. [11] investigated the effect of process parameters such as cutting speed, feed, drill point angle, and chisel edge on drilling of glass fibre-reinforced polymers. The experiments were designed according to the L9 orthogonal array. The ANOVA results revealed that feed rate is a significant parameter.

Experiments must be planned in such a way that they can be carried out at a low cost. Taguchi methodology is a commonly used method for designing experiments where one can infer the maximum amount of information from minimum input. ANOVA is a method used to estimate how input parameters influence output parameters. Several authors have previously designed experiments using the Taguchi method [12]. Shunmugesh et al. [4] conducted ANOVA studies on drilling carbide fibre-reinforced composites and found that the feed rate was the most important factor, while the drill tool material was the least important.

Derlin is a commercial homopolymer used in high-strength mechanical applications such as gears, protective equipment, medical devices, and delivery devices. Derlin is established as a commercial commodity that can be used as a substitute for metal. It has a higher wear resistance and a wide operating temperature range of $-40°C$ to $120°C$. It's compatible with other polymers and metals, and it's more dimensionally stable. It has a high tensile strength, improved stiffness, increased fatigue resistance, and improved impact resistance [11].

Data mining is a machine learning technique that may be efficiently utilized to analyze all system properties within models. It is amalgamation of machine learning, databases, and statistics. The use of DM in the manufacturing industry began in 1990 and has gradually gained the attention of the research community [13]. The various learning algorithms of data mining are (1) clustering, (2) classification, (3) association analysis, and (4) regression analysis [14]. Among these classification methods is supervised machine learning, in which the derived model can be represented as If-Then rules or decision trees. These rules are straightforward and easy to understand, and they can be extended to new datasets if the classification rules' accuracy is sufficient. Motivated by the strength of the decision tree algorithm, it is used in this work to determine the most significant parameters and to extract information from the data.

Quinlan [15] created an algorithm for decision trees (DTs) named the Iterative Dichotomiser 3 (ID3). The later version of this is the enhanced version, C4.5. Balasundaram et al. [16] proposed a step-by-step procedure of applying a decision tree algorithm for two machine flow shop scheduling problems. It is applied for an L9 orthogonal array dataset for surface roughness and roundness error. The decision tree algorithm requires the output variable to be categorical. Below-average values are considered fine, and above-average values are considered high when converting numerical values of surface roughness and roundness error. The root node is the attribute with the most information, and its contribution is primarily for influencing output variables.

According to a comprehensive literature review, there are no works related to Derlin material drilling and output analysis using data science. Drilling Derlin material with process parameters like feed, speed, and pecking was the first step. Surface roughness and roundness were the performance characteristics that were assessed. Finally, the influence of input parameters on output parameters was investigated using ANOVA and the decision tree algorithm.

7.2 MATERIALS AND METHODS

Derlin was chosen as the work piece for drilling and is shown in Figure 7.1(a), and the drilling was carried out using a FANUC Computer Numerical Control C Model horizontal drilling machine. Drilling was done with HSS drill bits and hard carbide drill bits. The key reason for using two different drill bits is to investigate the effects of each. Table 7.1 shows the experiment's input process parameters and their levels. Figure 7.1(a) shows the actual Derlin material. Figures 7.1(b) and 7.1(c) show the Derlin material after drilling by two drill bits. Surface roughness and roundness error were the output responses that were measured. The experiments were planned according to the L9 orthogonal array. A coordinate measurement machine was used to measure the roundness error of the drilling, while a surface roughness tester was used to measure the surface roughness error. Each reading was taken four times, and the average of the results was reported. ANOVA and a decision tree algorithm were used to study how input process parameters affect the output parameters. The schematic diagram of the experimental setup is shown in Figure 7.2.

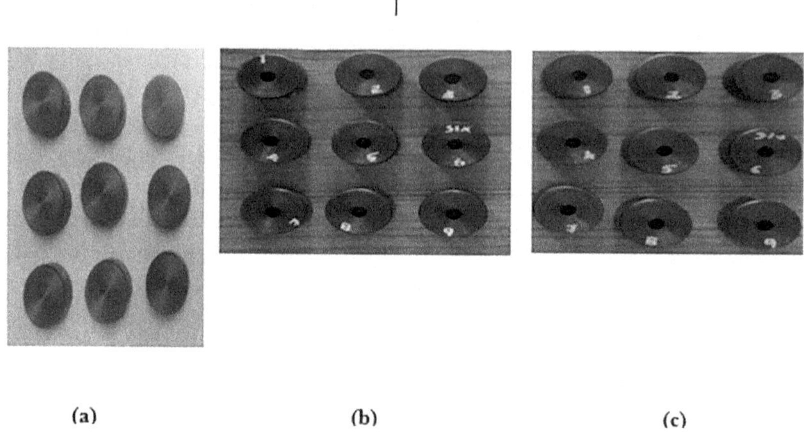

(a)   (b)   (c)

FIGURE 7.1   Selected material and drilled samples: (a) Derlin material, (b) drilled Derlin material with hard carbide drill bit, and (c) drilled Derlin material with HSS drill bit.

TABLE: 7.1

Process Parameters and Their Levels

Levels	Spindle Speed (N) (RPM)	Feed (f) (mm/rev)	Pecking (mm)
1	500	0.05	1.5
2	750	0.10	3.0
3	1000	0.15	4.5

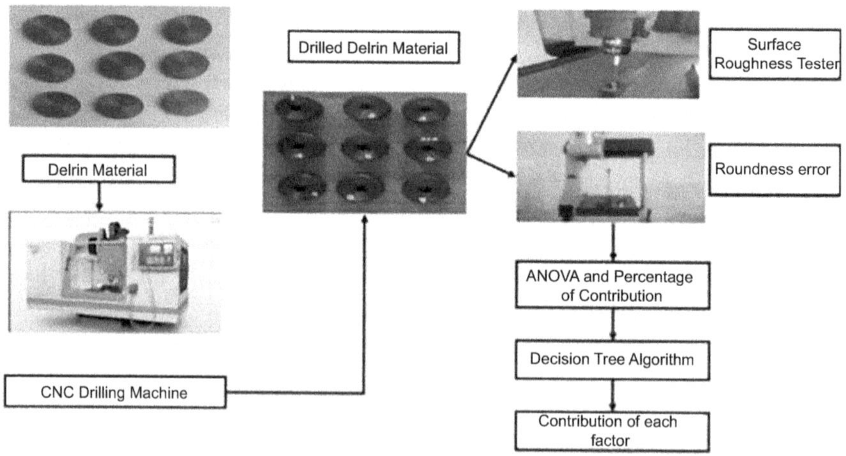

FIGURE 7.2   Schematic setup of work.

TABLE 7.2
L9 Orthogonal Array for Experimentation

| | Input Parameters | | | Output Parameters | | | |
| | | | | Hard Carbide Drill Bit | | HSS Drill Bit | |
Sl. No.	Speed (rpm)	Feed (mm/ rev)	Pecking (mm)	Surface Roughness (micron)	Roundness Error (micron)	Surface Roughness (micron)	Roundness Error (micron)
1	500	0.05	1.5	0.4545	0.006	0.857	0.015
2	500	0.10	3.0	0.255	0.003	0.796	0.026
3	500	0.15	4.5	0.376	0.008	0.604	0.031
4	750	0.05	3.0	0.432	0.003	0.506	0.038
5	750	0.10	4.5	0.298	0.007	0.534	0.025
6	750	0.15	1.5	0.356	0.009	0.598	0.01
7	1000	0.05	4.5	0.578	0.004	0.701	0.055
8	1000	0.10	1.5	0.428	0.006	0.921	0.021
9	1000	0.15	3.0	0.402	0.005	0.709	0.035

7.3   RESULTS AND DISCUSSION

The experiments were conducted using the L9 orthogonal array, and the results are listed in Table 7.2. The following section discusses the output parameters for both tools. The process parameters were also studied using ANOVA and data science.

7.3.1   SURFACE ROUGHNESS OF HARD CARBIDE DRILL BIT

Figure 7.3 depicts the main effect plots of the surface roughness of the solid carbide tool with respect to speed, feed, and pecking. There is an increase in surface values when there is an increase in speed. At higher speeds, the tool's interaction with the work piece occurs at a faster rate, generating more heat at the drill's surface, resulting in an increase in surface roughness. At a feed rate of 0.10 mm/min, the surface roughness was the lowest. Lower feed rates need more time to drill, resulting in the formation of more cracks and burrs on the surface and an increase in surface roughness. When the feed rate was set to 0.15 mm/min, the drilling tool travel time was extremely short, resulting in more defects on the surface and an increase in surface roughness. Pecking is the process of plunging a drill tool into a workpiece and then retracting it to the surface. Surface roughness values are lower when pecking for 3.0 mm when compared to other values. When pecking was at a higher level, the time it took for the tool to feed down, in, and out of the tool was longer, resulting in an increase in surface roughness.

The ANOVA values of the surface roughness of the hard carbide drill bit are shown in Table 7.3. From Table 7.3, it can be inferred that feed (57.86%) was the dominant factor for surface roughness. Feed plays a vital role, as it decided the

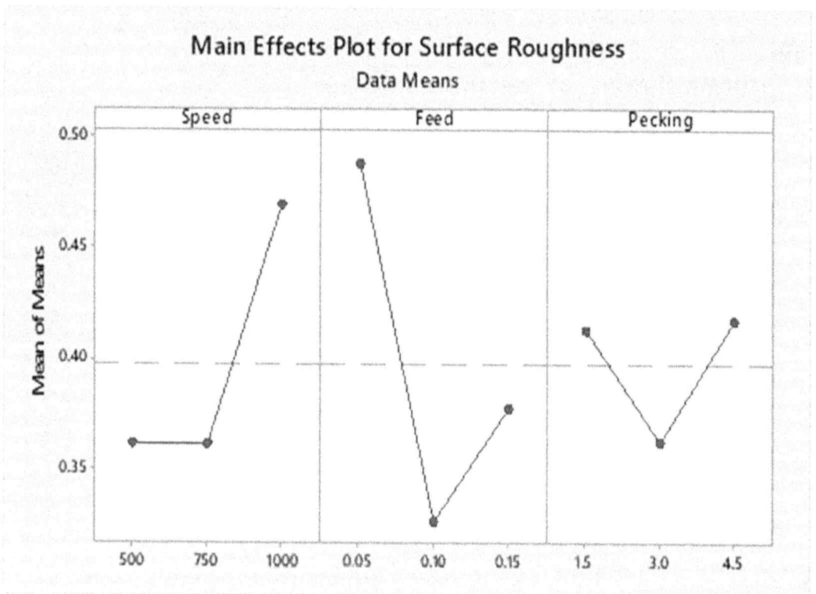

FIGURE 7.3   Main effect plot for surface roughness for hard carbide drill bit.

TABLE 7.3

ANOVA for Surface Roughness Using Hard Carbide Drill Bit

Source	DF	Adj SS	Adj MS	$F_{calculated}$	Percentage of Contribution (%)
Speed	2	0.023077	0.011538	20.68	32.79
Feed	2	0.040712	0.020356	36.49	57.86
Pecking	2	0.005456	0.002728	4.89	7.75
Error	2	0.001116	0.000558		1.6
Total	8	0.070361			

amount of drilling tool travel time during the drilling process. Following feed, speed (32.79%) and pecking (7.75%) were the influencing factors.

The DT algorithm of the surface roughness of the hard carbide drill bit is shown in Figure 7.4.

The rules are:

IF speed > 750 then surface roughness is poor
IF speed ≤ 750 and feed ≤ 0.05 then surface roughness is poor, else good

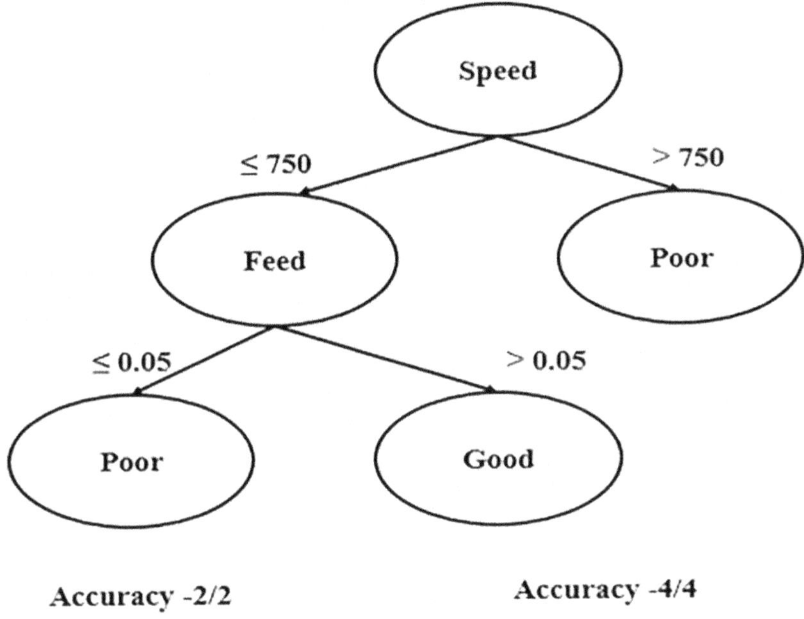

FIGURE 7.4   DT algorithm of surface roughness of hard carbide drill bit.

These rules classified all the instances in the L9 orthogonal array. Their accuracy was 100%. From the experiments, the contribution of speed was 32.77%, feed was 57.82%, and pecking was 7.74%. From the decision tree, it is understood that speed was most significant parameter, followed by feed for drilling using the hard carbide tool bit. From the results, it is understood that speed and feed contributed significantly, and pecking had less contribution on surface roughness using the hard carbide tool. The same thing was confirmed in the decision tree algorithm. The information gain of speed and feed was significant, whereas pecking had a very minimal effect.

7.3.2 Surface Roughness of HSS Drill Bit

Figure 7.5 shows the main effect plots of surface roughness vs. speed, feed, and pecking. The surface roughness increased when the speed was set to 500 rpm. Drilling was not uniform at lower speeds due to the higher strength of the Derlin material. When the speed was set to 1000 rpm, there was more generation of heat, which resulted in an increase in surface roughness. When compared to other factors, the surface finish was better when the speed was set at 750 rpm. It can be seen that when the feed rate is less, there is less surface roughness. When the feed rate is lower, the drilling time is longer, resulting in better surface roughness. When the feed rate is increased, the time taken for travel of the drill tool is more, which creates more uneven surfaces, resulting in more surface roughness.

The ANOVA values of the surface roughness of the HSS drill bit are shown in Table 7.4. From Table 7.4, it can be inferred that speed (57.83%) was the dominating factor in the case of the HSS drill bit. Pecking (30.03%) was the next dominating factor, followed by feed (11.58%). Thus, it can be concluded that optimal selection of speed was the dominating factor in the case of the HSS drill bit. The DT algorithm for the surface roughness of HSS drill bit is shown in Figure 7.6. The rules are:

IF speed > 750, then surface roughness is poor
IF speed > 500 & ≤ 750, then surface roughness is poor; else good

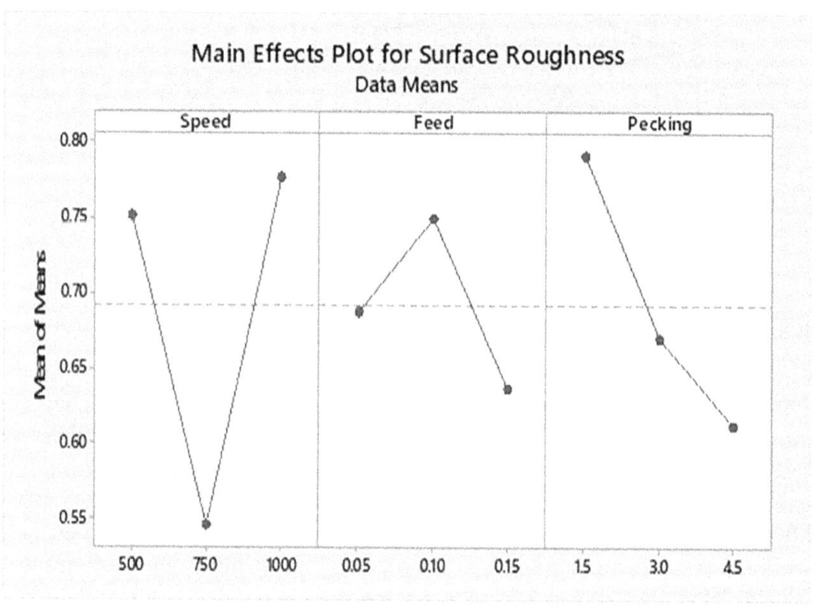

FIGURE 7.5   Main effect plot of surface roughness using HSS drill bit.

TABLE 7.4

ANOVA Values of Surface Roughness of HSS Drill Bit

Source	DF	Adj SS	Adj MS	F-Value	Percentage of Contribution (%)
Speed	2	0.096543	0.048271	97.83	57.83
Feed	2	0.019331	0.009665	19.59	11.58
Pecking	2	0.050131	0.025065	50.83	30.03
Error	2	0.000987	0.000493		0.56
Total	8	0.16692			

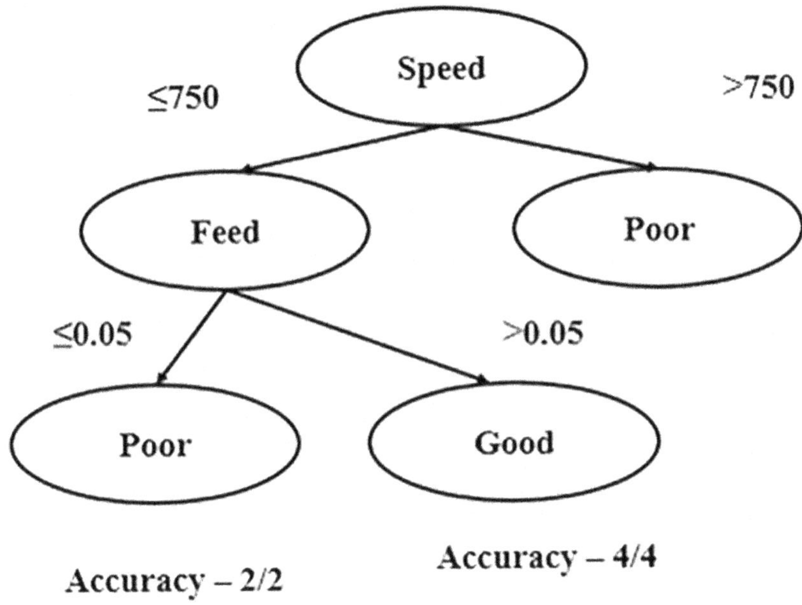

FIGURE 7.6   DT algorithm for surface roughness of HSS drill bit.

These rules are categorized into eight instances inside the L9 orthogonal array. Its accuracy is 8/9 which is 88.89%. From the experiments, the contribution of speed is 57.81%, feed is 11.57%, and pecking is 30.02%. From the decision tree, it is understood that speed is the most significant parameter. From the results, it is understood that speed and pecking contribute considerably and feed has less contribution to surface roughness using the HSS drill bit, in the decision tree algorithm, speed is enough for data classification due to its substantial information gain relative to other parameters.

7.3.3   ANALYSIS OF ROUNDNESS ERROR USING HARD CARBIDE DRILL BIT

Figure 7.7 depicts the main effect plot of roundness error in relation to feed, speed, and pecking. Figure 7.7 also reveals that as the speed increases, there is an increase in roundness error up to 750 rpm and a decrease in roundness characteristics as the speed increases further. In terms of feed rate, the roundness error was highest at 0.15 mm/min. When the feed rate was increased, the tool travelled very quickly, which resulted in an improvement in roundness. When the tool travel was decreased, less heat was produced, resulting in a decrease in roundness values. In the case of pecking, 1.5 mm of pecking gave higher roundness values. The tool movement into drilling was less, and it travelled a shorter distance during each pass, which resulted in uniform drilling. Multiple passes of the drilling tool led to an improvement in roundness of the drilled surface.

The ANOVA for roundness error using the hard carbide drill bit is shown in Table 7.5. From the table, it can be seen that pecking is the most dominant factor, followed by feed. Speed was the least dominating factor in the case of roundness. Thus, the factor contributing the most to the roundness error is pecking in the case of the hard carbide drill bit.

The DT algorithm of roundness error using the hard carbide drill bit is shown in Figure 7.8.

The rules are:

IF pecking is ≤ 1.5 mm, then roundness error is high
IF pecking is > 1.5 mm and ≤ 3 mm, roundness error is minimum; else high

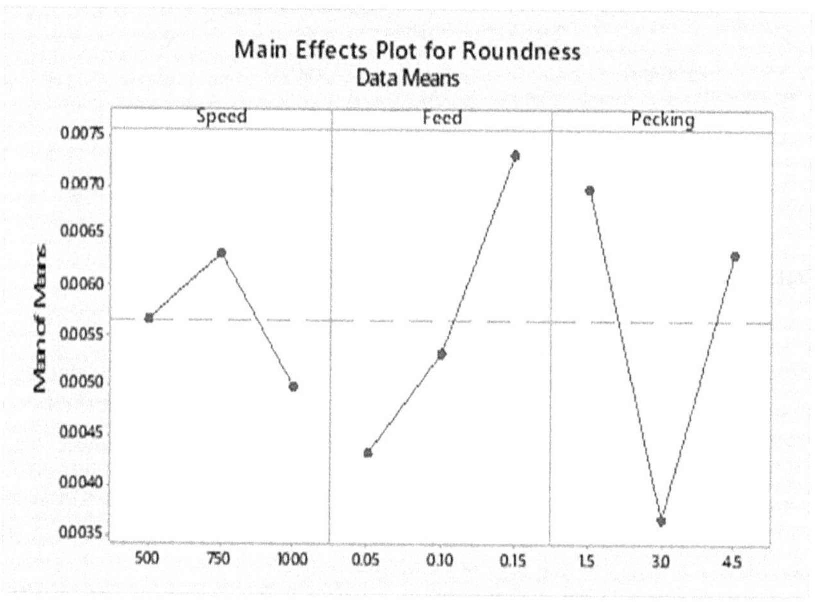

FIGURE 7.7 Main effect of roundness error for hard carbide drill bit.

TABLE 7.5

ANOVA of Roundness Error Using Hard Carbide Drill Bit

Source	DF	Adj SS	Adj MS	$F_{calculated}$	Percentage of Contribution (%)
Speed	2	0.000003	0.000001	4.0	8.33
Feed	2	0.000014	0.000007	21.0	38.88
Pecking	2	0.000019	0.000009	28.0	52.77
Error	2	0.000001	0.00001		0.02
Total	8	0.000036			

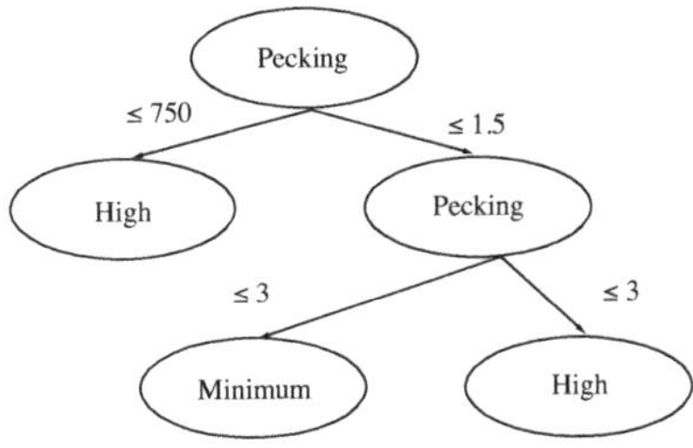

FIGURE 7.8 DT algorithm for roundness error using hard carbide drill bit.

These rules are categorized into eight instances inside the L9 orthogonal array. Its accuracy is 8/9 which is 88.89%. From the experiments, the contribution of speed is 8.10%, feed is 37.83%, and pecking is 51.35%. From the decision tree, it is understood that pecking is most significant parameter. From the results, it is understood that pecking and feed contribute considerably, and speed has less of a contribution on roundness error using the hard carbide tool bit, whereas in the decision tree, algorithm pecking is sufficient to classify the data, since it has high information gain as compared to other parameters.

7.4 ANALYSIS OF ROUNDNESS ERROR USING HSS TOOL

Figure 7.9 depicts the main effect plots for roundness error. In the case of an HSS drill bit, the roundness error increases as the drilling speed increases, as shown in

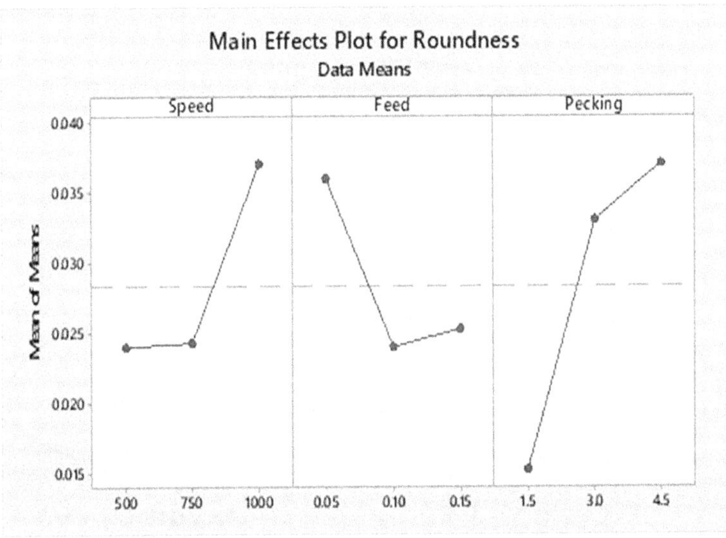

FIGURE 7.9 Main effect plots of surface roughness using HSS tool.

Figure 7.9. The size of the hole is not uniform when the speed is increased during the drilling process because more material is removed at a faster rate. As the feed rate is taken into account, the roundness error decreases up to the middle level and increases as the feed rate is increased further. The tool travel time is very short at higher feed rates, resulting in uneven hole formation. In the case of pecking, the roundness increases when the pecking is at a higher level. When the tool is plunged at a slower rate repeatedly, the formation of the hole is uniform, which results in a decrease in roundness characteristics.

The ANOVA for the roundness tool is shown in Table 7.6. It can be inferred that speed, feed, and pecking were the influencing factors in the case of roundness error. Pecking was the dominating factor by 55.35%, followed by speed (22.91%) and feed (18.05%). Thus, for roundness, the pecking of the drill tool must be minimum, which ensures that the hole formed is uniform. The DT algorithm of the surface roughness using the HSS tool is shown in Figure 7.10.

TABLE 7.6

ANOVA of Roundness Error Using HSS Tool

Source	DF	Adj SS	Adj MS	F-Value	Percentage of Contribution (%)
Speed	2	0.000330	0.000165	6.15	22.91
Feed	2	0.000260	0.000130	4.85	18.05
Pecking	2	0.000798	0.000399	14.89	55.41
Error	2	0.000054	0.000027		3.63
Total	8	0.001440			

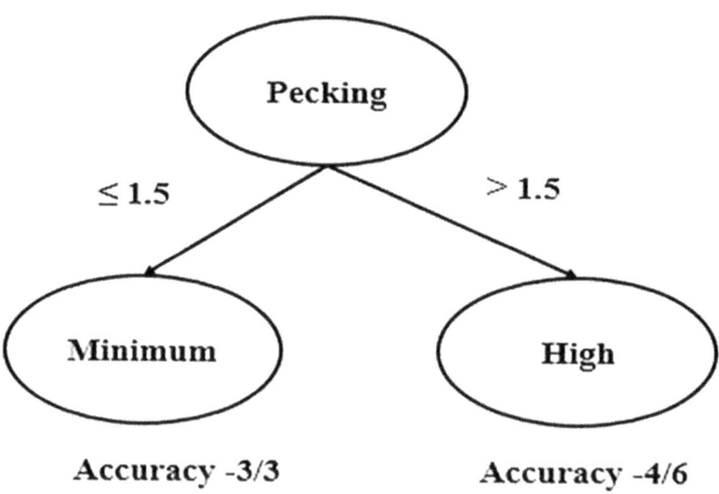

FIGURE 7.10   DT algorithm roundness error using HSS tool.

IF pecking is ≤ 1.5 mm, then roundness error is minimum; else high. These rules are categorized into eight instances inside the L9 orthogonal array. Its accuracy is 8/9 which is −88.89%. From the experiments, the contribution of speed is 22.82%, feed is 18.03%, and pecking is 55.35%. From the decision tree, it is understood that pecking is most significant parameter. From the results, it is understood that pecking contributes considerably and speed and feed have less contribution to roundness error using HSS. The decision tree algorithm also confirms that pecking is the significant parameter to classify the data.

7.5 CONCLUSION

In the present work, drilling of Derlin material was carried out. Experiments were designed according to the L9 orthogonal array. ANOVA was utilized to study the effect of input parameters over the output parameters. Data mining algorithms are used in the present work. Hence, in this work, an effort is made to propose a set of rules in the form of If–Then rules on a machining dataset for Derlin with an L9 orthogonal array. These rules are simple to understand for semi-skilled labourers, and they can classify or predict future data without depending on technical expertise. To apply the decision algorithm, Derlin is drilled using both a hard carbide drill bit and HSS tool bits. From the experiments, the following observations are made.

1. For SR using hard carbide, the contribution of speed is 32.77%, feed is 57.88%, and pecking is 7.77%. From DT, the speed and feed parameters are sufficient to classify the dataset, and pecking has no effect. It is also observed that if feed and speed are low, the surface roughness is poor. For moderate speed and feed, SR is good. Similarly, if speed is high, whatever the feed, its SR is poor. For SR using HSS, the contributions of speed are 57.81%, pecking is 30.2%, and feed is 11.57%, whereas in DT, speed alone is sufficient to classify the dataset. For high and low speed, SR is poor, and for moderate speed, it is good.
2. For roundness error using hard carbide, the contribution of speed is 8.18%, feed is 37.83%, and pecking is 51.35%. In decision trees, pecking is the sole predominant element for classifying the dataset. For high and low pecking, the roundness error is high, and for medium, it is minimum.
3. For roundness error using the HSS tool, the contribution of speed is 22.88%, feed is 18.03%, and pecking is 55.33%. From DT, pecking is sufficient to classify the dataset. For low pecking, the roundness error is minimum and when high depth of pecking is considered roundness error is high.
4. Due to the graphical representation of results from the decision tree algorithm, it is easy to understand the rules and the rules and can leverage considerable expertise. In the future, it could be used for multi objective optimization problems.

REFERENCES

1. Daniel, D. Jafrey, and K. Panneerselvam. "Mechanical and thermal behaviour of polypropylene/Cloisite 30B/Elvaloy AC 3427 nanocomposites processed by melt intercalation method." *Transactions of the Indian Institute of Metals* 70 (2017): 1131–1138. https://doi.org/10.1007/s12666-016-0908-6

2. Sakthi Balan, G., M. Ravichandran, and V. Santhosh Kumar. "Study of ageing effect on mechanical properties of Prosopis juliflora fibre reinforced palm seed powder filled polymer composite." *Australian Journal of Mechanical Engineering* 18 (2020): 1–13. https://doi.org/10.1080/14484846.2020.1806194

3. Ahmad, Furkan, Ankit Manral, and Pramendra Kumar Bajpai. 'Machining of thermoplastic composites." *Processing of Green Composites* (2019): 107–123. https://doi.org/10.1007/978-981-13-6019-0_8

4. Shunmugesh, K., and K. Panneerselvam. "Multi-response optimization in drilling of carbon fiber reinforced polymer using artificial neural network correlated to meta-heuristics algorithm." *Procedia Technology* 25 (2016): 955–962. https://doi.org/10.1016/j.protcy.2016.08.187

5. Shunmugesh, K., and K. Panneerselvam. "Grey relational analysis based optimization of multiple responses in drilling of carbon fiber-epoxy composites." *Materials Today: Proceedings* 4, no. 2 (2017): 2861–2870 https://doi.org/10.1016/j.matpr.2017.02.166

6. Palanikumar, K. "Application of Taguchi and response surface methodologies for surface roughness in machining glass fiber reinforced plastics by PCD tooling." *The International Journal of Advanced Manufacturing Technology* 36 (2008): 19–27. https://doi.org/10.1007/s00170-006-0811-0

7. Prakash, S., J. Lilly Mercy, Manoj Kumar Salugu, and K. S. M. Vineeth. "Optimization of drilling characteristics using grey relational analysis (GRA) in medium density fiber board (MDF)." *Materials Today: Proceedings* 2, no. 4–5 (2015): 1541–1551. https://doi.org/10.1016/j.matpr.2015.07.080

8. Díaz-Álvarez, Antonio, Ángel Rubio-López, Carlos Santiuste, and María Henar Miguélez. "Experimental analysis of drilling induced damage in biocomposites." *Textile Research Journal* 88, no. 22 (2018): 2544–2558. https://doi.org/10.1177/00405175177251

9. Bouhrour, Nesrine, Peter H. Nibbering, and Farida Bendali. "Medical device-associated biofilm infections and multidrug-resistant pathogens." *Pathogens* 13, no. 5 (2024): 393. https://doi.org/10.3390/pathogens13050393

10. Vinay, Naveen Prakash, Rakshith Gowda, Nithyananda, Ranjith, and Srikantamurthy "The effect of process parameters on quality characteristics in the drilling of aluminium–metal matrix composites." *Engineering Proceedings* 59, no. 1 (2023): 53. https://doi.org/10.3390/engproc2023059053

11. Kaviarasan, V., R. Venkatesan, and Elango Natarajan. "Prediction of surface quality and optimization of process parameters in drilling of Derlin using neural network." *Progress in Rubber, Plastics and Recycling Technology* 35, no. 3 (2019): 149–169. https://doi.org/10.1177/147776061985507

12. James Dhilip, Jafrey Daniel, J. Jeevan, D. Arulkirubakaran, and M. Ramesh. "Investigation and optimization of parameters for hard turning of OHNS steel." *Materials and Manufacturing Processes* 35, no. 10 (2020): 1113–1119. https://doi.org/10.1080/10426914.2020.1765254

13. Djellouli, A., K. Benyelloul, H. Aourag, S. Bekhechi, A. Adjadj, Y. Bouhadda, and O. ElKedim. "A datamining approach to classify, select and predict the formation enthalpy for intermetallic compound hydrides." *International Journal of Hydrogen Energy* 43, no. 41 (2018): 19111–19120. https://doi.org/10.1016/j.ijhydene.2018.08.122

14. Saidi, F., N. Sebaa, A. Mahmoudi, H. Aourag, G. Merad, and M. Dergal. "Structural electronic and mechanical properties of YM2 (M= Mn, Fe, Co) laves phase compounds: first principle calculations analyzed with datamining approach." *Solid State Communications* 274 (2018): 9–20. https://doi.org/10.1016/j.ssc.2018.02.013

15. Quinlan, J. Ross. "Induction of decision trees." *Machine Learning* 1 (1986): 81–106. https://doi.org/10.1007/BF00116251

16. Balasundaram, R., N. Baskar, and R. Siva Sankar. "A new approach to generate dispatching rules for two machine flow shop scheduling using data mining." *Procedia Engineering* 38 (2012): 238–245. https://doi.org/10.1016/j.proeng.2012.06.031

Index

For Product Safety Concerns and Information please contact our EU
representative GPSR@taylorandfrancis.com
Taylor & Francis Verlag GmbH, Kaufingerstraße 24, 80331 München, Germany

www.ingramcontent.com/pod-product-compliance
Ingram Content Group UK Ltd.
Pitfield, Milton Keynes, MK11 3LW, UK
UKHW022312100726
473146UK00009B/437